OCCULTISM AND MODERN SCIENCE

by

T. Konstantin Oesterreich

qp

QUID PRO BOOKS

New Orleans, Louisiana

Published in 2012 by Quid Pro Books, as a *Digitally Remastered Book.*™

ISBN 978-1-61027-915-4 (paperback)

QUID PRO BOOKS
Quid Pro, LLC
5860 Citrus Blvd., Suite D-101
New Orleans, Louisiana 70123
www.quidprobooks.com.

δεῖ ἐλευ θέριον εἶναι τῇ γνώμῃ τὸν μέλλοντα φιλοσοφεῖν

An open mind is the gate-way to philosophy.
—Plato

OCCULTISM AND MODERN SCIENCE

BY

T. KONSTANTIN OESTERREICH

PROFESSOR OF PHILOSOPHY AT THE UNIVERSITY OF TUBINGEN

TRANSLATED

FROM THE SECOND GERMAN EDITION

METHUEN & CO. LTD.
36 ESSEX STREET W.C.
LONDON

PREFACE TO THE FIRST EDITION

THE present book is concerned with a field of knowledge which is not much cultivated in Germany, but which has for a number of years been academically recognized by the English-speaking and the Latin races. At the publisher's request it is addressed to the general public, including therein such of my colleagues as have not been more closely concerned with the subjects dealt with. I have tried to describe from a non-partisan point of view the scientific position as it seems to me to exist. That this position is as yet far from clear in many respects will be obvious as I proceed. But it follows that it is obvious that we have to do to a considerable extent with a new field of knowledge which is not yet ripe, and which German science is called upon to join in cultivating, so that at last certainty may be reached as to what is actually true and the proper philosophical consequences deduced therefrom.

ORSELINA, LOCARNO
September, 1920

PREFACE TO THE SECOND EDITION

THIS book, the first edition of which was exhausted in six months, has accomplished what I hoped it would, and directed the interest of persons capable of scientific thought to the problems of parapsychology. At last there is movement in many places, though it need hardly be said that in many others dogmatic slumber still prevails. The reception of the book by the general public also goes to prove that we live in a time of mental freedom, which is not governed by dogma, and should be ready to make a great advance, if it were not that the general conditions of life make German scientific work nowadays difficult even in this field.

The theoretical interpretation which I have attempted of the facts recorded must for the present remain in details hypothetical, and my meaning would be wholly misunderstood if the various lines of thought which I have developed were taken to be positive dogmatic conclusions.

TÜBINGEN

June, 1921

CONTENTS

CHAPTER   PAGE

      INTRODUCTION   -   -   -   I

I.   HELENE SMITH—STATES OF IMPERSONA-
        TION   -   -   -   -   17

II.   MRS. PIPER AND PSYCHOMETRY   -   37

III.   CROSS-CORRESPONDENCE   -   -   59

IV.   EUSAPIA PALLADINO—TELEKINESIS   -   71

V.   EVA C.—PROCESSES OF MATERIALIZATION   99

VI.   THEOSOPHY—RUDOLF STEINER   -   129

      CONCLUSION. THE SCOPE FOR NEW
        INVESTIGATIONS   -   -   -   154

      LITERARY APPENDIX -   -   -   170

OCCULTISM AND MODERN SCIENCE

INTRODUCTION

CIVILIZATION (Kultur), nowadays, is undergoing a series of critical transformations. We daily witness political changes, the gravity and completeness of which surpass any events known to history. For the first time, all nations are linked up into a causal whole, and the world has become one single political unit. The international means of exchange—enhanced by the increased facilities in material and intellectual communication which was the result of the nineteenth century— have brought to an end former conditions under which conflicts only broke out between adjoining countries.

During the world war, nearly all the civilized nations, together with the majority of their dependent peoples, were drawn into combat. All the seas and the greater part of the surface of the continent became a theatre of war. But while the fires of this general conflagration still smoulder, a new catastrophe of international character threatens to overwhelm the globe.

Russia, which only two years ago showed every indication of utter collapse and appeared to have dwindled to a mere shadow of itself, has now spread civil war and wholesale terrorism throughout the rest of the world.

The downfall of civilization which took place in Russia is in danger of engendering universal collapse, throwing back the civilized nations in particular to more primitive conditions than those which characterized ancient history. The situation, however, is much more serious to-day than when the invasion of the German hordes heralded the advent of the middle ages with its retarded civilization. For at that period, the invading tribes had no definite object in view beyond youthful enthusiasm for war, conquest and adventure—whereas Russian Bolshevism is intent on destroying the very framework of " Kultur," as well as the structures of society— an intention carried through with ruthless determination wherever Bolshevism takes root.

But while the foundations of mankind sway under the impact of this torrent, and none can safely predict whether he may not be engulfed by this aftermath of the world war, we visualize on the horizon of pure thought phenomena such as precede the end of a cosmos. The red flush of a setting sun already casts its dying reflection over the whole body of knowledge of the modern world. All is changed. The scientific work of the three last centuries has been proved one-sided and incomplete. It originated with the study of phenomena of movement in Nature, initiated by Copernicus, Kepler, Galileo, and Newton ; but these were the product of dead matter, and formed but part of a system which cannot be taken as representing the whole.

The widening of our intellectual horizon and the consideration of other branches of reality have already produced far-reaching alterations in our mechanical conception of the world. They have brought about revolutionary changes in theory, even in the realm of dead matter. It is enough to mention in proof of this the dissolution of the elements and the principle of relativity. But far more fundamental are the changes wrought in our conception of the universe by reason of the introduction of facts from the mental and organic worlds. These two sections represent independent spheres of study. When they are given due weight to in our theories, the world of our conception assumes a still more changed appearance.

The advance of the analysis of the life of the mind has made clear the essential difference of its behaviour from that of matter. All the important deductions of modern psychology during the last two centuries have but widened the breach between psychology and natural science. Towards the end of the nineteenth century, the

world lived under the delusion that the principles
of a fundamental science—that of elementary
mental phenomena—had been discovered in
experimental psychology. It was assumed that
this science could be used for the study of the
mental world in the same way that mechanics
are employed to investigate the conformation of
inorganic bodies. This strange delusion is over.
No serious student would now allow that experimental psychology, operating within the
sphere of the science of mind, can even approxi-
mately be compared to that of mechanics, as
considered in relation to the natural science.
All hope of discovering laws of like structure in
the life of the soul, to those that control the
domain of mechanics, has remained unfulfilled.

No lasting factor exists in mental life—apart
from the centre of consciousness to which we
ascribe all mental acts and conditions—whereas
all the mechanics of natural science presuppose
the existence and continuity of their elementary
physical component parts. In the life of the
soul no law of conservation of energy exists
similar to that which governs the laws of
mechanics. Herbart's attempt to establish a
mechanics of the life of the soul on the assumption
that perceptions were separate elements con-
stituting mental matter, has met with no greater
success when repeated by experimental psychology
on a more advanced scale. If we wish to make a

true comparison between the material world and the world of mind, we find that the material elements do not correspond one by one to separate individual moments in the life of the mind, but rather that each individual psyché taken separately as a whole in itself corresponds to a separate material element. On the other hand, however, while mechanical laws alone govern the relationship between the separate material elements, all the so-called psychological laws are based on the distinctions within the individual soul—each separate mind being for itself a separate mental universe.

But the state of psychology, together with that of the mind itself, becomes further complicated through the fact that all minds—so far as we know—are in close communication with the physical world. Natural science is concerned with objects which represent an independent sphere of reality (or at least can be treated as such). The experiences of the mind are corollaries of events in the material world. We take each mind to be allied to a physical organism. It gets into touch with the outer world through the medium of the senses, and it is only cognizant of the existence of other minds through its perception of foreign bodies connected with these other souls. On the other hand, the mind is able—partly by its own conscious action though mainly unconsciously—to modify its own organism, thus conveying to other minds signs of its own existence. The whole of the action of the mind, as shown by its effect upon the material world, entirely destroys and shatters the mechanical conception of the world, which believed itself entitled to consider consciousness as quite isolated—without outward influence of any kind upon material things. The theory of parallelism was a desperate attempt to establish this isolation without admitting the influence of consciousness on the physical sphere.[1]

Of an equally fundamental character is the upsetting of the mechanical conception of the universe which results when we consider the construction of organisms which take their material from the physical world. Whether this process is attributed to the act of God or to vital forces of a special kind—whether one hypostatizes the unconscious functions of the individual mind—or constructs special laws for organic life—in every and any case the facts cannot be explained unless some new factor is presupposed.

Modern psychology and New Biology have joined together to uproot the older materialistic conception of the world, though they have not succeeded in replacing it by any other definition. It is but too evident that the development of the organic world and the appearance and dis-

[1] Theism does not admit that the Material world represents an independent sphere.

appearance of minds from the ken of our immediate experience remain wholly unexplained and unexplainable. The world of experience presents us with the picture of a cosmos, in which ever new factors become active, only to vanish into nothingness, leaving as little traces of their disappearance as of their previous inception.

If the actual world of experience is to be regarded as the whole of Reality, nothing is of more frequent occurrence than the *creatio ex nihilo* and its antithesis—entire annihilation.

But the crisis in present-day views of the universe has not yet reached its final stage. On the contrary, we are confronted with the prospect of a much more serious upheaval, which will result in a new conception of the universe. A still further revolution is once again to widen the horizons of thought by bringing into consideration hitherto unnoted realms of reality.

There are volcanic signs of disruption under the superficial layer of official culture. As a matter of fact, conceptions which are in clear contradistinction to science have always continued to exist. Superstitions of every sort have remained ingrained in civilized nations; magic and witchcraft exercise their influence as of yore, only to be opposed by the thin veneer of education. It is, however, indisputable that the situation at present indicates change to no inconsiderable degree.

Mid the lumber of superstition and delusions of every sort there are occasional psychic and psychophysical phenomena of a peculiar nature of their own which has made them the real foundation of, or rather the focus round which, phantasies formerly eluding scientific research have crystallized. They are phenomena of so strange a nature, that they are qualified to influence in decisive fashion our entire conception of the universe, and even of life itself. The whole subject is summed up in the word : OCCULTISM—an unfortunate appellation, for it is not only things called " occult " which are mysterious. We designate as mysterious anything which is not clear to us, and of which it is not easy to find the explanation. And as a matter of fact very few things *are* perfectly clear ; all the real facts of life have their mysterious side—even the simplest and those of daily occurrence. We do not know actually why a stone falls to the ground, why a pane of glass breaks when hit, or how it is that we see when the light-rays reach our eyes. And when we ascribe these events to a cause—as formerly to the pressure or shock of the ether-atoms—or, as at present, to the circulation of electrons round a positive nucleus—we have again to accept another fact, just as inexplicable, in place of the one last established. For we do not know why this is so— why electrons exist and move in this or that definite way, producing this or that effect. The child is right when it demands an explanation for every new fact brought to its notice. We are often forced to answer " It is so," and cannot help ourselves except by forbidding any further questions. In spite of this, the child is right to question us. His thirst for information is not yet blunted by a continuous repetition of events. For, so far as we are concerned, it is nothing less than intellectual laziness when we think that all that we experience, all that we know, has become clear and unenigmatical. The grown-up has merely become accustomed to facts and questions no longer. Everyday happenings have lost for him the sense of the mysterious, the unintelligible. It takes something unexpected and new to make him desire an explanation. Every great discovery in the realm of natural science therefore fills him with feelings of awe and mystery. Among these can be classed the X-rays ; the Hertzian waves ; the radioactive emanations, and numerous others. But after a while they all enter the category of natural phenomena. The same applies to " occult " problems. Only so long as telepathy, clairvoyance and materializations appear as something out of the ordinary, are they surrounded by an aura of mystery.

Swedenborg, who communicated with the spirits as though they were still in the flesh (as he believed), found nothing out of the way even in this. In the same way the relationship of the innumerable automatic writers of the present day to the spirits with which they claim to be in communication (through the Planchette) is also treated by them as quite in the ordinary course of things.

The debated facts of occultism are only different from many others in the world because of their comparative rarity, but this does not mean that they are unique in their nature, for it is possible to arrange them in groups or classes. They may be compared to works of a creative nature in art or science, which also like occult facts cannot be produced artificially, but can only be observed after they have come into existence. The appellation " occultism " is consequently extremely inapposite. Despite this, it may be accepted provisionally as an accepted designation as many other words are accepted, if it is understood that it does not possess any other significance except the conventional designation of a certain definite sphere of problematic matter. After all, the actual word used is of no importance.

In recent times the word " scientific " has been added to " occultism," and by this is meant scientific research into the problematic realm of the facts under discussion. A better and more trenchant term is " Parapsychology "— or also as Richet puts it, " Metapsychology." It stands to reason that the most competent investigator for the major part of this domain should be the psychologist, an assertion beyond dispute. With regard to the remaining part of the facts to be investigated, the question might well be raised as to whether naturalists and biologists are not the best investigators ; though as one delves deeper into the subject, it becomes obvious that this work, too, cannot be accomplished without the psychologist, and that what is really required is collaboration.

What makes Occultism so repellent to the professional scientist is the mental milieu in which the problematic facts under observation are so often presented. The parapsychologic problems exercise a peculiarly fascinating influence on all half-educated individuals, whose inherited conceptions of the universe have been uprooted, but who long, nevertheless, for more complete knowledge. They are drawn to them as moths to the light of arc-lamps. In the case of decided hysterical or neuropathological types, they soon come to believe that they are in mysterious contact with the transcendental world, and they then develop a peculiar spiritual fanaticism which makes all discussion with them as hopeless as it is unrefreshing. They look down on science with ineffable disdain, meet every critical objection with instinctive enmity, filled as they are with the secret fear that they may find themselves to have been mistaken.

This general mental predisposition tends to make them the easy victims of astoundingly impudent frauds. I have noticed with amazement how women who were both educated and intelligent allowed themselves to be duped in the most transparent fashion by a woman medium at a spiritualistic séance. So crass were the means employed, that I could not resist the temptation of competing with the medium, and in my turn giving an exhibition of the same methods, in order to prove the absolute lack of critical analysis reached by these women in their search for the miraculous.

But even the individuals under examination—the mediums—are not infrequently of similar mental constitution. It is possibly a mistake to assume that all mediums are hysterical, although many of them are. But, after all, with hardly an exception, they take their stand on their spiritistic convictions. In dealing with them, therefore, tact is required to an unusual degree. The psychical researcher has, furthermore, to adapt himself to strange and often quite repellent opinions. This is essential. Mediums are extremely sensitive in regard to a sceptical attitude towards spiritualism. The experience of Flournoy with Helene Smith is a case in point, though his book made her famous all over the world. Their connection was broken off after it had lasted for several years simply because he could not commit himself to countenance the phenomena she presented by the desired spiritistic explanation. I myself have tried in vain to maintain my relations with an automatic writer. These too came to an abrupt end when I, though admitting the fact of automatic writing, refused to agree to its spiritistic interpretation. But even when the investigator manifests the utmost care and caution, he is not immune from disillusion and unpleasant surprises. And yet despite all this it is impossible any longer to refuse to discuss the problems of occultism.

A large part of the more serious occult literature is contributed by authors who have devoted themselves exclusively to this domain. We approach their work with a natural scepticism and reserve, for the absence of any other scientific productions on their part deprives us of other standards by which to judge the quality of their parapsychological investigations or the mental value of their occult publications. In some cases the impression left on our minds is such that we cannot refuse them credence, though in others we do not get beyond a *nonliquet*. The more curious and astounding the result of their deductions, the more we are inclined to reserve our judgment, even in the event of a general favourable impression of a given work.

However, for some time past, parapsychological investigations have not been exclusively confined to convinced occultists.   To-day there is a considerable increase in the number of competent investigators who have proved their ability in other departments before turning their attention to Occultism.   This already started in the eighties and nineties of last century. Those of the older generation have vivid recollections of the sensation when the founder of Astralphysics, C. F. Zoellner, his friend the originator of Psychophysics, G. Th. Fechner, and the distinguished English Physicist, Crookes, occupied themselves with the mediums Slade and Home, all three affirming the reality of phenomena hitherto regarded as definitely refuted by all physical experiments.   Although it was possible to contend successfully against Zoellner, Fechner and Crookes, that they placed too much faith in their mediums, and did not take sufficient precautions against possible deception, this contention holds less and less with regard to more recent investigators, more especially as it has become the rule to keep the possibility of fraud well in mind.   Formerly the whole problem could be waived aside.   Zoellner could be called insane (and also erroneously accused of having committed suicide), and Crookes met with no better treatment.   But is it really admissible to accuse every fresh student of being half-witted or unscientific, for the simple reason that having taken up Occultism, and remaining equally competent in his own sphere, he deduces certain definite though quite abnormal facts from his later study ?   Surely the probability is greater, that their unanimity expresses the real state of affairs, and that Occultism in consequence really deals with new facts of a peculiar character.

True, we are not dealing with happenings that can be observed at any time anywhere.   They are peculiar to certain persons, to be made use of wherever they are to be found.   But in psychology this situation is not rare.   It has already been mentioned, that all creative work which is of any importance in the history of intellectual development has been confined to certain individuals only, and in their case even has not at all times been in evidence.

The scepticism brought to bear on such subjects in Germany at the present day goes much too far.   Those familiar with foreign literature are forced to the conclusion that it is simply based on ignorance of the information already available.   It is characteristic enough that until a few years ago the chief periodical on the subject, containing the most important and essential work and data, the " Proceedings of the Society for Psychical Research," was (so far as I know) only procurable at the Munich State Library (though now it can also be found in the Berlin State Library).   We are merely behind the times in this respect, and this attitude is a reversion to Materialism, and not the only one at that. The present state of affairs is more and more unworthy of German science, and my object in this book is to put an end to it.

The best and shortest way is to give a summary of the results of the examination of several contemporary mediums. For that purpose I have chosen those who demonstrate the phenomena under consideration in a peculiarly distinct and clear-cut fashion. These are the Swiss medium, Helene Smith ; the American, Mrs. Piper ; and the Italian, Eusabia Palladino—the three most famous mediums of modern times. No others have been so thoroughly and continuously examined as they, for they remained for years in succession under scientific observation. Once acquainted with the facts established through their mediumship, we shall be in possession of the best proofs by which to judge Occultism as a whole. We shall, however, not content ourselves with the examination of these three special cases, but occupy ourselves—either in conjunction with them or separately—with other more recent cases.

CHAPTER 1

HELENE SMITH—STATES OF IMPERSONATION

I WILL begin with Helene Smith, a medium through whom some forms of parapsychic phenomena are demonstrated with particular clarity, though many or the most characteristic of them are only found occasionally.

Helene Smith is the pseudonym of a mediumistically inclined Swiss lady, whose acquaintance was made by Th. Flournoy, Professor of Philosophy at the University of Geneva in the winter of 1894. She was then thirty years of age, of middle-class origin—an employee in a business-house in Geneva, where, thanks to her intelligence, her position improved gradually. At the age of 14 her first abnormal experiences began. They took the form of nocturnal apparitions. Later on, new phenomena manifested themselves; and finally, after she had joined spiritistic circles in 1892, and had herself become one of their disciples, she developed into a regular medium, experiencing while awake phenomena previously only experienced by her at night. Thus, in her writing, written characters differing from her own began to appear here and there. To these were added, later, acoustic phenomena and rapping. Finally the somnambulistic dream stage was attained. These conditions recurred with increasing frequency, thus rendering Helene Smith one of the most remarkable subjects for analytic observation in connection with the so-called " Phenomena of Impersonation "—the modern form of " Possession." At such times her mind seemed to leave her body altogether, and to be replaced by another. So at least it appeared.

For five years, 1894–1898, Flournoy was able to observe her in the course of innumerable séances, during which period she derived no material benefit from the proceedings, and received no pecuniary remuneration in her capacity of medium. Nor were the greater part of her performances of a character to invite deception. She laid no claim to prophecy, nor did she excel in the usual telepathic or paramnestic performances—to say nothing of giving any special demonstrations in the way of physical phenomena.

Inspirational phenomena predominated. These were shown with the greatest ease and abundance, and there is no other medium, through whom under equally close observation such phenomena were obtained with greater frequency.

The spirits which apparently declared themselves through Helene Smith fall into two categories—historical and non-historical personages; or, rather, in order to avoid any preconceived theory, those with which it is possible to connect a prototype in history, and those with which it is not. The former are confined to a few cases, such as Victor Hugo, Leopold Cagliostro, the famous magician of the eighteenth century, and Marie Antoinette. The latter is infinitely the richer group. These spirits did not all declare themselves through Helene Smith at the same time. On the contrary ; her medial life may be divided into varying periods, or, more correctly speaking, cycles. Victor Hugo was the first to appear. His impersonation lasted five months, and then gradually diminished. His place was increasingly taken by Leopold Cagliostro, of whom a visual apparition proceeded the actual impersonation. As a matter of fact, it should be noted that in the case of Helene Smith there existed in general some peculiar connexion between visual and inspirational phenomena which has still to be explained in detail. From now onwards, Cagliostro became Helene Smith's actual " control." He appears to have been present—according to the observations made by Flournoy—throughout her entire further development. The impression received is that he never leaves her—he is conversant with her whole life. His conversation and writing, including the orthography, appear to date from the eighteenth century, and his physiognomy bears distinct resemblance to a historical portrait of Cagliostro. We have here a case where the physiognomy of the medium is subjected to considerable changes during the state of trance, taking on a resemblance to the features of the spirit which may happen to appear to be in control of the medium—a very telling proof of the magnitude of the transformation in the mind of the medium while in a state of impersonation. All her impersonations have this characteristic. Flournoy describes again and again the ineffable art with which Helene Smith portrays the character of the moment.

" Helene should be seen when the ' royal ' trance is full and complete : grace, elegance, distinction, at times majesty in pose and gesture—the actual demeanour of a queen. The most subtle shades of expression—charming amiability, queenly condescension —indifference and withering contempt, are shown in rapid succession on her countenance and bearing as the défilé of her courtiers pass before her in her dream. The play of her hands with a real handkerchief and fictitious appurtenances—fan, lorgnette, smelling-salts with a screw-top in a little bag attached to her girdle—her curtseys, her movements full of careless grace, as she never omits to throw back her imaginary train at every step—every thing, every smallest detail is perfectly and naturally worked out.

Seen under these conditions, she must impress us as a finished actress, except that, in contradistinction to the real artist, she is entirely merged in her rôle, and retains no consciousness of her own real personality. On the other hand, however, these conditions cannot be accepted as proof of the conclusion that real impersonation through the spirits which have passed over actually exists.

" Possession," by Leopold Cagliostro, betrayed certain imperfections only too clearly. Helene Smith has no knowledge of Italian, neither had her Leopold ! Moreover, comparison of the two handwritings proved that there was no resemblance between that of the pseudo-Cagliostro and the real one ; but that it was simply and solely the distorted handwriting of Helene Smith.

It is not different with the impersonation of Marie Antoinette. The old-fashioned orthography and handwriting are also shown in this case, but so are the same defects. The real handwriting of the French Queen was different, and her accent was German and not English. It is equally significant that Helene Smith's Marie Antoinette should use such modern words as tramway and photography, of which the historical queen could not have heard. These circumstances prove that there can be no question of actual " possession," however finished the imitation of strange personalities may otherwise be. To this must be added the fact that there are stages of transition between Helene Smith and those ostensibly impersonated. At certain moments Helene Smith feels herself becoming Cagliostro. Her own normal existence is merged in that of the foreign personality which at that moment begins to live and stir within her mind and consciousness. She feels herself at one and the same time to be Helene Smith and Cagliostro; just as the poet, in moments of inspiration, may feel himself at one with the creatures of his brain. This would seem impossible, had the real Cagliostro taken possession of Helene Smith's organism. Still more convincing proof of the non-identity of these personages presumably impersonated in Helene Smith's trances is the occasional appearance of characters from novels which she had read. They too lay claim to be taken seriously.

In the cases quoted so far, we have only been concerned with copies of characters which Helene Smith had met in fiction or in historical tradition. But her imagination was capable of far greater flights. Whole cycles of fictitious persons and situations were evolved by her. They all bear an exotic stamp, headed by an Indian cycle. Just as the impersonation of Leopold was preceded by his apparition, so did the Indian cycle begin with the visions of Indian landscapes, and was followed by Helene Smith's impersonation of visionary Indian figures—apparently semi-historical personalities of the fourteenth century. From now on impersonations and visions began to intermingle in odd fashion. At the very moment that Helene Smith is metamorphosized into an Indian woman, she visualizes a little monkey in hallucination and plays with it.

The Mars-cycle is still more imaginative. Here, too, there was an intermingling of visions and impersonations. Helene Smith " acts " the inhabitants of Mars, and at the same time visualizes the imagined countryside, houses, plants, etc., of Mars. This cycle is further characterized by the introduction of a " Martian " language. The Martians impersonated speak no language known on earth—only Martian; and their written characters differ entirely from any " earthly " alphabet. Flournoy has examined both languages and script most minutely. The very melodious speech—vowel sounds being estimated at seventy-three per cent— and e's and i's preponderating—was proved by him to be a thoroughly grammatically-grounded language; but no independent tongue, merely a somewhat transformed French. The language was a specious production, constructed and used with amazing skill, but nevertheless a faked transformation of a European tongue; in fact, no new independent language. It was really wonderful how Helene Smith used this speech, which was evidently invented by her, with no opportunity of practising it out of her trances. It is as remarkable as if some one read through a foreign grammar, and forthwith began to speak in that language. This is the only comparison which can be made to Helene's power of using the newly invented script and of writing it fluently. Invention and mastery followed close upon one another, to the apparent exclusion of the usual practice necessary under normal circumstances.

Her abnormal memory and facility of reproduction when in a trance has been established through two other facts. One day Helene Smith wrote in entirely unknown characters. Investigation proved these to be Arabian letters, and after further observation they were recognized as copied from a dedication, which a Geneva physician had written several years previously in his book, " En Cabylie." It was at least six years since Helene had seen that particular volume. Characteristically enough, these were the only Arabian letters ever traced by her hand. The second proof of her trance hypermnesie is demonstrated by her use of several genuine Sanscrit words when presumably speaking " Sanscrit " in her trance. Closer investigation disclosed the fact that she had formerly held spiritualistic séances at the house of a person who dabbled in Sanscrit, where she may possibly have seen a Sanscrit grammar.

Given such prodigious feats of memory, Helene Smith's invented language and alphabet can well be understood. When the true nature of her trances was established beyond doubt, Flournoy made a bold move, and told the " Control " of the medium, Leopold Cagliostro, the truth to his face. The latter withdrew from the contest with the pointless phrase, " *Il y a des choses plus extraordinaires.*" The incident, however, did not end here ; but it had further consequences for Helene Smith. New cycles were evolved in attempts to outbid the previous ones, with a view to giving a still more striking proof of the genuine character of the impersonations. In the place of Mars and the Martians appeared Asteroid and Uranus, to which later was appended a Moon cycle. New languages and newer and still more phantastic alphabets were evolved.

We have here, therefore, clear evidence of an effort to create belief in an imaginary intercourse with a higher world. Have we the right to accuse Helene Smith of deception ? It does not seem so. In any case the question does not arise in connexion with Helene Smith in her waking state, in her normal everyday surroundings, but can only refer to her state in somnambulistic trance.

Helene Smith, in a waking state, can only be held responsible for such actions to the same extent as a person who commits them in a somnambulistic trance. But in the latter case responsibility is not usually attributed. We do not say that a person is responsible merely because he or she acts in a certain way, unless he or she has also complete control at the time of the normal mental faculties, and is not in an anomalous psychic state. Matters, however, are further complicated with regard to Helene Smith by the fact that her strange somnambulistic condition as well as her visions themselves were *both* occasioned by an impulse to prove the truth of her intercourse with supernatural spheres. All the same, Helene Smith cannot be held responsible for this when she is awake. Or are we to accept the explanation that Helene, while in a normal frame of mind, lends herself sub-consciously to such considerations ? Given such a hypothesis, even so she should not be held responsible, for there is no apparent sense in making a person accountable for thoughts and actions of which she is unconscious, while giving vent to them.

Helene Smith did not always enter into a complete trance ; the abnormal psychic processes often only manifest themselves in a semi-somnam-bulistic or even still vaguer state. Automatic writing was frequently produced, representing as it does, possibly, the most oft-recurring phenomenon of mediumship of the present day. We are here concerned with a most remarkable phenomenon, though so recurrent a one that it can be produced at any moment in spiritualistic circles. Its explanation alone remains problematic.

Completely developed automatic writing consists—according to the mediums—in that their hands write purely mechanically, without knowledge on their part of what they write. According to accounts by such authors as P. Janet and A. Binet, cases have occurred in which it was possible to converse with the mediums without disturbing them in the least in their writing. Cases have even been known in which the medium wrote with both hands simultaneously on different topics. There is no point in questioning these assertions as assertion of mere fact. But all cases are not of equally well-marked character. On the contrary, we find every possible grade and transition, ranging from normal voluntary active writing to these extremes.

According to the classic Anglo-French theory attributed to Taine, we are dealing with " dissociated psychic processes," or, when another ego appears to express itself in writing, with " secondary personalities." This theory is in complete harmony with that of the conception of the mind by Wundt, which until lately practically dominated German Psychology. According to this theory the individual ego is no more than a synthesis of separate psychic parts, and there is no permanent element which can be called the soul. According to this view mediums possess two or more of such mental syntheses, whereas in the normal human organism all mental processes are united in the form of one single unit. This conception is irreconcilable with that of the modern monadic view of the mind propounded by me in my " Phenomenology of the Soul." My view, however, is also able to deal with the automatic processes of mediumistic phenomena without having to resort to the hypothesis of purely physiological " Reflex Phenomena." This moreover is a hypothesis which cannot be sustained, for automatic documents are often so completely coherent that something more than reflex action must be presupposed. To my mind there are only two possibilities : either the writing-motions of the medium are controlled by intellectual activities and thoughts, which do not reach his conscious apperception, and of which he is as oblivious as we are oblivious of words addressed to us when we are otherwise engrossed, or else the medium is at fault when he professes unconsciousness of such thoughts. This again invokes a dual possibility : either the medium forthwith forgets what has been thought or written, or merely imagines that he does not know what he is writing about. The latter supposition, strange and improbable though it appears, is not unlikely. Innumerable cases have been established through the examination

of Psychaesthenics, where the latter complained
that they wrote absolutely mechanically and did
not know what they wrote, despite proof on
investigation that they were entirely aware of it.
It seems that such persons lack the normal
complement of emotional feeling in intellectual
as well as in other directions. And inasmuch as
they are painfully conscious of their deficient
sensibility, so that their own sensations appear
to themselves either strange or non-existent, it
is possible that a similar result in a more marked
manner may take place in intellectual matters,
particularly as the process of thought are in
themselves so unsubstantial. It is, therefore,
hardly surprising if they feel as though they had
lost the power of independent thought. From
this the further deduction might be made that,
when the mediums write automatically, the
emotional consciousness which accompanies
thought is suspended, and they are not therefore
conscious of thinking. This explanation may
apply to some cases and sufficiently explain them ;
though it fails when the " personality " who has
been writing automatically is incapable, when
asked, of giving the sense of what has been so
written. Upon this view it is impossible to
determine experimentally whether retrograde
amnesia supervened, or whether the act of
thought in question did not penetrate to the
conscious apperception ; for the effect is the

same in either case. The individual concerned cannot indicate the contents of what was written either during or after the act of writing.

It is quite wrong from the monadic point of view to speak of " Subconsciousness." We speak of consciousness in all cases where we are cognizant of the psychic processes within us. " Subconsciousness," however, would imply that we possess such knowledge " below " our consciousness. This would mean that at one and the same time we have and have not such knowledge, and consequently contains a complete inner contradiction. We may assume the existence of as many subconscious processes as we please, but we must not talk of " subconscious consciousness."

Automatic writing presents no specific qualities. It may consist of meaningless lines ; on the other hand, it may also have a quite coherent meaning, either commonplace or full of interest. Poems of considerable beauty have even been evolved in this fashion. With Helene Smith the automatic writing appears to be derived from the same functional psychic combination as that which fills the whole consciousness in her somnambulistic impersonations.

But such automatic phenomena are not only confined to mediums. The best known case in connexion with a non-medium is that of Miss Beauchamp, examined by Morton Prince. In this instance, too, we are concerned with a whole crowd of so-called " personalities " who purport to be independent subjects, but who are, without doubt, only special modifications of Miss Beauchamp. Their relationship to each other is the same as that of the spirits to the medium, and they too make their presence manifest—even in Miss Beauchamp's presumably normal state—by means of automatic writing, hallucinations, etc. Here too, as in the case of Helene Smith, it might be argued that these strange phenomena are the result of suggestions made by the investigator himself; but at any rate, they do exist, and are worthy of more minute study; at least, as much as cases of hysterical anæsthesia, which it may be are also sometimes provoked by the treatment itself and caused by an intentional misdirection of the attention.

Such phenomena are, for the greater part, closely connected with the action of suggestion in hypnotic cases, the precise psychological nature of which we still do not know. The only experimental scientific German work which shows understanding of, and seeks connexion with, these problems is to be found in Ach's book on " Willing Experiments."

Apart from the phenomena of impersonation, Helene Smith seems to have evolved, though in a lesser degree, other phenomena of supernormal character in a narrower sense. She seems to have the gift of telepathy. Thanks to this gift, she is said to have had such an intimate knowledge of the private life of one of her fellow-clerks at her office that he was forced to give up his work there in consequence. On another occasion, for instance, she had a vision of Flournoy, said to be clairvoyance, when the latter was ill. However, these facts are less completely established than are the phenomena of impersonation, since they are necessarily based on the reports of the medium herself or on the testimony of others. Even so, they can not always be completely explained. For instance, one day the medium automatically put down the signature of a priest, who, it was discovered later, had lived in a small hamlet in Savoy in the beginning of the nineteenth century. This signature absolutely corresponded with the original as recorded on ancient documents. It was, however, not proved that Helene Smith had ever seen this signature, which had been entirely forgotten. It was only found possible to prove that she had once passed through the village. It is, therefore, impossible to say how this automatic copy of the signature should be regarded.

Equally vague in the case of Helene Smith are the reports on physical mediumistic phenomena. It was stated that a piano, violin and bell produced spontaneous sounds in her presence. Further, she informed Flournoy that once, after the visit of a man who was unsympathetic to her, a couple of oranges lying on the piano precipitated themselves in his direction immediately on his departure as the expression of her distaste of his presence. But the sole witnesses for the truth of this story are Helene Smith and her mother; we are therefore unable to gauge the probabilities of this assertion.

Finally, we should like to discover the exact importance of her mediumistic faculties to Helene Smith's general existence. Did they impair or advance the course of her life? This question is answered by the fact that the chief phenomena only took place during spiritistic séances, consequently only with assent of the Medium. It is true that she was unable to produce them at her will in the ordinary sense of the word, but Helene Smith created or refused to create the atmosphere in which they spontaneously developed. The only disturbances of her normal life were to be found in occasional semi-somnambulism, insignificant hallucinations, and illusions of compulsion. It also occasionally happened that an ordinary letter was replaced by one of her invented characters in the midst of her usual handwriting, or that the distinctive script of Marie Antoinette suddenly appeared altogether in its stead. Such occasional, though insignificant impediments to her normal mental existence were counter-balanced by many helpful

3

mediumistic phenomena. Among these must be reckoned the useful counsels of Leopold Cagliostro, conveyed by automatic writing, acoustic hallucinations, and other means. He advised her on her health, and his recommendations, concerning her participation in spiritualistic séances, invariably proved sensible and correct. This relationship between Helene and Leopold is nothing out of the common ; indeed it is of quite usual occurrence between the mediums who write automatically and the spirits which ostensibly control them. The mediums are in closest sympathy with the spirits, who constitute themselves their most faithful and intimate friends, to whom the mediums refer in all the great and small happenings of life ; and it is not uncommon for them to receive really valuable advice and instruction in this guise. This mental mechanism does not differ essentially from that of normal human beings. When we ponder over certain matters, the most valuable thoughts are sometimes conveyed to us passively in the form of inspiration. The mental life of the medium is characterized by the fact that these same inspirations appear at once in the glorified presentment of thought and advice, emanating from a different personality. In the case of Helene Smith these thoughts are presented through automatic writing and acoustic hallucinations. We are obliged to assume that the medium does not possess an immediate apperceptive consciousness of these mental happenings but is only apprised of them through automatic writing or hallucinations. It is impossible meanwhile to determine the reason why such mental processes, which do not even reach apperception, should so easily set the writing-muscles in motion, or produce hallucinations. Sometimes the " unconscious " processes in Helene Smith lead to more varied phenomena. One day as she was in the act of reaching from a cupboard an object too heavy for her physical strength, her arm suddenly stiffened, and Cagliostro explained later that he had been responsible for this to prevent her health from being impaired.

In other cases, when she wished to remember something, the answer presented itself in hallucinatory form. She recovered a brooch in a similar fashion, which she had lost one Sunday during a country walk. Again, on another occasion, she was assisted by her mediumistic faculties in placing the order of a client who wrote to ask for a material No. 13459. No one in the whole house knew to what he referred, neither did Helene Smith, who also hoped to find it. Suddenly she placed her hand quite mechanically on a roll of material, and when she looked closer she found it bore the number required. In all such instances her mediumistic faculties were of great use to her. But the greatest advantage gained thereby lay in their ultimately freeing her from her cramped material position. A rich American lady, who took a fancy to her, conducted her to a bank one day, and there made certain financial arrangements, which secured her independence for life. Her firm conviction that she was ordained for something better than that of a warehouse clerk was thus realized. She at once severed her connexion with her employers, but at the same time unfortunately with Flournoy, whom she had unsuccessfully tried to convert to spiritualism. Nothing further is known about her later medial development, though, considering how restricted Geneva circles are, inquiries on this point should not be difficult.[1] It is greatly to be desired—in view of the interest presented by the observation of the psychic development of every Medium, that this omission should be rectified at the earliest opportunity.

[1] Note by translator, Professor Oesterreich is misinformed. She took to psychic (inspirational) painting. *Cf.* "Annales des Sciences psychiques," *passim* up to 1914.

CHAPTER II

MRS. PIPER AND PSYCHOMETRY

THE case of the medium, Helene Smith, did not present any special problem in its main manifestations. The strictly supernormal phenomena were not sufficiently frequent to be either understood or admitted. It is a different matter when we come to the American medium, Mrs. Piper. Mrs. Piper, at an earlier date than Helene Smith, produced supernormal phenomena with such regularity and under such unimpeachable conditions that they can, with the greatest probability, be regarded as established facts. For decades she was under scientific observation and the result never varied. Thus we have here a case of which the supernormal character is above all suspicion. It is therefore no longer a question of the problem of the existence of supernormality; the problem lies in the ways and manners of its evolution.

Mrs. Piper was a married woman of the Boston middle-classes. Her scientific discoverer was W. James, whose attention was drawn to her in 1885. The manner of his discovery was both unromantic and unscientific in character; in fact, it sounds more like an old wives' tale. James's sister-in-law told him one day of an unknown woman who had been able to give details about the writer of an Italian letter, who was a stranger to her, simply by placing the letter on her forehead. James was sceptical, but sufficiently curious and interested to look Mrs. Piper up for himself. At her first sitting, however, his former suspicious attitude was replaced by the conviction that Mrs. Piper was producing supernormal psychic phenomena. In her trance she was able to give detailed information regarding James's relatives, though none of these lived in the neighbourhood. Some had settled in California, others in Maine, some were already dead.

She knew that one of James's children was dead. "Your child," said the spirit which claimed to be talking through Mrs. Piper to James, "has a playfellow here in our world, a boy named Robert Fr——"; and this name was found to be the actual name of a child who had died. James himself believed the information to be incorrect, and that the child referred to had been a little girl. Inquiries proved, however, that it was not the spirit which was wrong, but James—it *was* a boy. The medium made correct assertions about James—" You have just killed a grey-white cat by means of ether." James's mother-in-law lost a cheque book; it was found through her indications.

Curiously enough, though he visited Mrs.
Piper several times, James did not undertake any
further personal examination of the medium for
quite a time, though he kept himself permanently
informed about her through his friend Hodgson,
the Secretary of the American Society for Psychical Research. The latter, except when away
on a journey, instituted regular sittings several
times weekly over a period of twenty years, from
the date of his arrival in America (1887) until
his death in 1906. He carried out this duty,
if not always in the best of tempers, yet with the
utmost conscientiousness and in a most business-
like fashion. He also undertook to introduce to
Mrs. Piper all the visitors who came to see her,
as she considered it to be her religious duty to
place her strange gift at the service of science.

Shorthand reports of the automatic statements
of the medium were taken down at innumerable
sittings. In later years automatic writing came
to the fore, and voluminous records emanating
from the supposed spirits were the result. In
1900 the proceedings of the Society for Psychical
Research had already published 1,500 pages on
Mrs. Piper alone, of which half was devoted to
the minutes of her sittings. Since then, further
comprehensive publications have been printed,
so that the present material in hand comprises
some 3,200 pages, though many records of the
sittings and inspirational writings have not been

printed. To these must be added other material
on the subject published elsewhere. No other
medium, with the exception of Eusapia Palladino,
has been examined so often.

As time went on, Mrs. Piper's mediumistic
faculties became fainter. She found increasing
difficulty in falling into a trance, and indeed this became impossible after the middle of 1911, owing,
possibly to the after-effects of a shock experienced
at certain experimental sittings with the psychologist, Stanley Hall, and his assistant, Miss Tanner.
Despite this, she was still able to produce automatic writing in a state of normal consciousness.

The manner in which the supernormal phenomena manifested themselves through Mrs. Piper
during a lifetime is common to all mediums.
She fell into a trance, and the spirits then spoke
through her. At least so it happened at first,
though, later on, automatic writing was more
prevalent. But even so, for many years Mrs.
Piper remained in a state of trance when writing.
In order to attain this state at first she merely
held some one's hand ; after a few minutes'
spasmodic movements set in, resembling a slight
epileptic attack, with a lessening of the cutaneous
sensibility. Then came the state of impersonation. Mrs. Piper apparently withdrew from her
organism, and other individualities took her
place. Their number is a very considerable one,
certain of them recurring often, Phinuit,

George Pelham, Rector, Imperator, and others;
and later; after their death; Myers and
Hodgson. The "controls," or impersonating
spirits, however, were not always themselves the
communicators on their own behalf. It not
infrequently happened that one or the other of
the impersonating spirits volunteered the information that other spirits were present who said
certain things ("communicators"). The life-
like resemblance of those impersonated must,
according to the reports, have been unusually
strong; character, voice and demeanour were
almost uncannily accurate.

Nevertheless, Mrs. Piper is not distinguished
by any specific peculiarities from numerous other
mediums of lesser qualifications. The interest of
her manifestations rests in the first place on the
knowledge shown by the impersonated individuals, or rather by Mrs. Piper in the impersonation trance. This knowledge was not infrequently of supernormal nature—not so much as
regards its subject as in the manner in which
that Mrs. Piper was able to obtain it.

The character of supernormal phenomena
always remained the same. When Mrs. Piper in
trance was apparently controlled by certain
personalities, she often gave information concerning the name, character and past of those present
as well of others known to them, either alive or
dead. These details were always quite unin-

teresting : the description of some one's cane,
what sort of cuff-links he wore, and from whom
he had received them as a present, etc. She
made a point of reminding those present of
various little details of their past, of which she
was quite unlikely to have heard. Such information was mainly forthcoming when objects
belonging to those interested were placed before
Mrs. Piper (Psychometry). The knowledge
which she had of deceased persons was so astounding that many a sceptic, entirely opposed to
spiritualism, became a convert. The immediate
impression produced by the automatic writing
or conversation and the seemingly direct intercourse with spirits, who declared themselves by
questions and answers, were apparently much
more convincing than is the subsequent reading
of the minutes of the sittings. This explains
how it was that even an investigator of such
strength of mind as Hodgson, who originally
belonged entirely to the positivist school of
thought, was converted to spiritualism. His
friend, George Pelham, was apparently impersonated—after his recent death—in Mrs.
Piper, and reminded Hodgson of the varied details
of their former philosophical conversations.
Further, the " spirit " greeted all his old acquaintances. His parents were presented to him
under an assumed name ; but in vain, for he
recognized them nevertheless. All this made

such a profound impression on Hodgson that he came to the conclusion that the spiritistic interpretation was justified. The craving to learn more about life after death from his own experience became overwhelming, and he is reported shortly before his death to have said that he could hardly contain his impatience : " I can hardly wait to die."

James had a similar experience after Hodgson's death, with the latter's impersonation. The resemblance to the deceased and the supernormal nature of the information volunteered was so great that " I felt a slight shiver down my spine, as though I really had been talking to my old friend." And he too, who had for years defended the anti-spiritistic standpoint against Hodgson, now no longer felt able to reject entirely the spiritist explanation.

There is no doubt that Mrs. Piper did not obtain her knowledge by normal methods. Those who have studied but a few passages of the shorthand notes of the minutes must be certain of this. True, it is dry reading (at least, for those who are not greatly interested in such problems), for the notes are very trite and ordinary for the main part. The chief interest invariably centres in the question how Mrs. Piper could have known of these intimate details. And even Stanley Hall, who apparently possesses the typical positivistic scepticism of the average experimental psychologist, admits that " the control seems to possess faculties that appear supernormal."

The problem no longer runs : " Do supernormal phenomena occur in the case of Mrs. Piper ? " but " Which hypothesis is the more likely to explain them ? "

Disciples of the spiritistic interpretation draw attention to further considerations. For instance, spirits who are impersonated in a medium soon after death make extremely confused statements, as if they had not yet completely found themselves. This is particularly noticeable with regard to the spirits of those who died from mental or similar diseases, tending to prove that they still bear traces of their mental deficiencies. It also happened that one of the " spirits," who was impersonated in Mrs. Piper, explained to a lady present at the séance, that he had just appeared to one of her relatives who had died immediately afterwards. It was proved later that the person in question had actually died and that the " spirit " had actually appeared to him shortly before death. The very words, which appear to have been heard by the dying man, were repeated by Mrs. Piper. And yet none of these arguments are incontrovertible. Every one of the evidential cases might be explained as an elaboration by the creative imagination of Mrs. Piper's telepathically acquired knowledge and by her telepathic faculty working in conjunction

with the minds of others—in the instance given
with that of the dying man. Without the
hypothesis of telepathy, all attempts at explanation are abortive. And in addition to the telepathic perception of the immediate actual mental
processes of those present at the séance we have
also to assume that the medium could read
thoughts which were latent. When Mrs. Piper
informs Professor James that he has just killed a
cat with ether, there is a possibility that he might
have given a casual thought to this fact at that
precise moment. When, on the other hand, she
gives him information about distant relatives or
a dead child the above theory appears improbable.
It must therefore be assumed that she was herself
able to reproduce even mere latent memories of
those present.

A special difficulty arises in those cases where
Mrs. Piper made correct statements in contradiction of the thought of the person who was
apparently the telepathic source of her information, that person making a mistake. This applies
to the case where Mrs. Piper indicated the correct
sex of the child, while James was wrong. If,
however, the origin of her assertions is to be
found in James's memory, it must be assumed that
there are, so to say, deeper strata of subconsciousness, otherwise her declaration would have
agreed with his erroneous opinion. That such
deeper strata do exist is proved by the fact that

under certain artificially induced conditions it is possible by narrowing the circle of consciousness to improve the memory and correct mistakes. Mrs. Piper's telepathic power seems to have gone direct to such latent memories.

But how is it that Mrs. Piper, when shown an object belonging to a person unknown to all those present, was yet able to give information about it later proved to have been correct ? Or when she disclosed matters even unknown to the latent memory to those present at the séance ? These psychometric manifestations have so far been considered inexplicable.

In order, however, to attempt to explain them, it has been assumed that all objects are surrounded, so to speak, by a psychic aura or by the " Life-spirit " of their owner. ("Influences.") Either conception, particularly the latter, is quite nebulous. The additional hypothesis deduced from them, that, for instance, it is not right to place articles which were the property of different owners close to each other, as they infect each other and give bad psychometric results, has not been verified. Stanley Hall on one occasion showed Mrs. Piper an object which was not the one originally chosen to be shown to her, but only bore a marked resemblance to the original. She was nevertheless enabled to make correct communications applicable to the owner of the original " real " object, in spite of the fact that it possessed neither " psychic aura," nor was it steeped in " nerve spirit." She had apparently been duped.

Another explanation lies in the assumption that the survival of personality is so limited that only shreds of memory are left in the world, and it is of these shreds that the mediums are able to take advantage. This conception implies an exceedingly strange misconception of the nature of the mind, as indeed of memory in particular, and results in a reversion to the ideas of Herbart. Just as Herbart materialized the individual acts of perception into permanent atoms, so upon this theory acts of memory are regarded as concrete facts : and this, though individual acts of memory, even when repeated with reference to the same object, cannot by any means be considered as identically the same. Furthermore this theory is at fault in assuming that if two different people remember the same event it must be the case of an identical remembrance. On the contrary, there would be two distinct acts of memory, as each person has his own individual memory, even though applied to the same event. If this hypothesis is to be adopted at all, it must be applied consistently and clearly. We shall then need an entirely new foundation for psychological theory. In exact opposition to the monadic conception of the soul, it will be necessary to assume that the psyche, like the body, is also composed of individual parts, capable either of a permanent, independent existence, or at any rate of a continued existence for a certain length of time. These separate parts may find themselves combined at certain times, just as the body is composed of atoms, which, if placed in a different juxtaposition to each other, would produce other bodies. Dismemberment and materialization would not only be true of the memory, but of all other mental phenomena. And the result would be—unless we go on to assume the existence of separate specimens of the same mental phenomenon—that we should have to say that to some degree different individuals are actually constituted from the same parts. For instance, a colour noticed by another person and myself would be actually the same identical perception in both of us. The same thing would apply to an emotion or manifestation of will power. For the theory, if it is right for memory, is right also for all other mental acts, and thus demonstrates its own absurdity.

In my opinion, all psychometric manifestations alike can be traced back to Telepathy, and this I have pointed out in my " Fundamental Notions of Parapsychology." It must be assumed that Mrs. Piper was in unbroken subconscious telepathic nexus with almost everybody, so that much of their actual experiences or memories was

telepathically transferred to her, and at her
disposal while she was in a trance and able to
recall it.   If this is so she would, on being shown
a watch, remember its owner, to whom certain
associations would necessarily be attached, in the
same way that we, on receiving a gift, may think
of the donor and possibly of his relatives or
other common acquaintances.   For this reason
I should like to suggest the term " Paramnesy "
or " Metamnesy " for psychometric phenomena.
The supposition that spurious spirits and not
Mrs. Piper are responsible for such communications would merely be an explanation created
by the imagination, and it is of daily occurrence
in modern occultism by reason of traditions and
beliefs which are passed on from one medium
to another.

That the spiritistic interpretation actually
presents difficulties to spiritists themselves has
been clearly proved by the recent attempt made
to explain Mrs. Piper's trance, not as genuine
impersonations, but as founded on a telepathic
nexus not only with the living but also with the
spirits of those who had passed over and were
continuing their existence transcendentally.   It
is true that it is not possible to refute this any
more than the usual spiritistic interpretation ;
but it is still true that all positive proof of spiritism
is unjustified, for whatever the communications
may be by which spirits prove their existence,

they must themselves be verified, in order that their validity may be accepted. But verification is only possible when the facts are vouched for by living people or proved by documents. And where this is possible, it is also possible in principle to ascribe the knowledge of the medium to Telepathy or Clairvoyance.

That Mrs. Piper was in possession of telepathic and possibly clairvoyant faculties also seems to be confirmed by various data. On one occasion a sitter was informed by her through a " spirit " that there was a defective place, a crack under a certain window in her (the sitter's) house. Another time it was directly arranged with one of the " spirits " (G. Pelham) that he should observe the doings of a certain person and report at the next séance what had happened in the meantime. This actually was done. The " spirit " reported that the person under observation, who lived far away in Washington, had taken a photograph to an artist on a certain day with a request to paint a portrait from it. This was quite correct. Not even the man's wife was aware of the incident.

There are, however, various positive considerations which militate against the spiritistic character of Mrs. Piper's state of trance. For instance, Dr. Phinuit, who lay claim to being a French doctor at the beginning of the nineteenth century, speaking through her, had no knowledge whatever of the medicines used at that time. Mrs. Piper's most outstanding failure lay, however, in being unable to communicate the contents of a letter, unknown to all who were then alive, left by a stranger who had died. The attempt was twice repeated, to fail in both instances. Once, too, there was the question of a letter left by Hodgson, which he had promised to communicate to his friends if possible after his death, through Mrs. Piper, as proof of his continued existence. Even though Hodgson was apparently impersonated soon afterwards in Mrs. Piper, his attempts to give the contents of the letter proved quite abortive. This led to the conclusion that Mrs. Piper's efforts as a whole were only connected with her telepathic nexus with the living; though possibly this conclusion may go too far. It is only certain, either that she is not in continuous telepathic nexus with everyone, or that her memories are not always entirely within her control, otherwise she would have received telepathic news of this letter during Hodgson's lifetime when it was written, and remembered it later. She was not gifted with equal paramnestic faculties with regard to all the sitters.

These abortive attempts also prove that Mrs. Piper was not always capable of clairvoyance, or she would have been able to decipher the letter by that means. And despite her supernormality there are other errors and gaps in her manifestations. At times, for instance, she gropes in doubt after a name, and occasionally does not get beyond similarities in sound. Thus Gibbons was pronounced as Kiblin, Giblin, and so forth. And the definite impression left is the same as that when we ourselves are half unable to recall a name, a result which is much in favour of the explanation already given in regard to " Psychometry." Half-true, inaccurate, and totally false communications have also been given ; as, for example, the wrong date of the delivery of the photograph to the painter in the episode above mentioned. In other cases, it was not possible to establish the exact truth.

But all these inaccuracies, defects and negative results cannot shake the positive material. Its wealth is overwhelming.

So far as I am aware, no one who came into personal contact with Mrs. Piper, or who was concerned in first-hand reports about her, had any doubt as to the supernormal nature of her mind, and her supernormality is as securely established as any historical event. It has been proved scientifically, and there can be no further discussion as to the fact. Most of the investigators fared just as James did. Those who grew up in the atmosphere of the departing nineteenth century necessarily brought scepticism and rationalistic prejudices to bear on the preliminary study of parapsychological problems, but the case

of Mrs. Piper could not in spite of all their
scepticism be lightly dismissed. In order to be
quite certain steps were taken to place her and
her relatives under continuous supervision by
detectives, and nothing in the least suspicious was
ever discovered. She was several times sent to
England, to stay as a guest in a private house,
in totally strange surroundings. Her luggage,
and practically the whole of her limited corre-
spondence—she never wrote more than three
letters a week—were examined, with equally
fruitless results. What paraphernalia she would
have needed had her demonstrations been founded
on fraud ! She was fully informed, so to speak,
on each person who came to her; and not only
on the person himself, but also on his friends and
relatives, both alive and dead. And as she never
knew who was likely to come to her, she should,
by rights, have possessed a register or family
record of everyone under the sun. Though even
the most comprehensive index would have been
useless without her supernormal faculties, for
she would not only have had to memorize this
index in its entirety, but also to identify each
visitor, even when, as repeatedly happened, he
was introduced to her under an assumed name.

As a matter of fact she only learnt of her
peculiar condition through the reports of third
parties. She had herself no recollection of her
trance. Unlike Helene Smith, she appears to

have had herself no consciousness of a dual personality when she was passing slowly from her normal condition to her trance state.

While Helene Smith became aware of her abnormal psychic processes through automatic writing and semi-somnambulistic conditions, Mrs. Piper was either in a complete state of impersonation and trance-somnambulism or entirely normal. There was no transition through intermediate stages. Consequently it was only through the testimony of others that she became aware of anything remarkable about herself. She personally was more inclined to the telepathic interpretations than to the spiritistic, and she is the only one of the three great mediums with whom we are concerned who shows definite reserve with regard to spiritism.

Bearing in mind the fanatic devotion evinced by such individuals towards spiritism on the whole, it is decidedly refreshing to come across a medium of such remarkable powers who adopted such a critical attitude : " My opinion is to-day (1901) as it was eighteen years ago. Spirits of the departed may have controlled me and they may not. I confess that I do not know."

However unusual the interest created in Mrs. Piper's case by the wealth, abundance and, above all, the careful scientific control to which it was so long subjected, her case does not stand alone. She has not only been rivalled by English-speaking

mediums, but by Germans also. Tischner reports on them, and the psychologist, Professor
Baensch, was repeatedly present at such experiments. In one instance, the latter himself
handed the medium, X, a small silver Turkish
coin, which could not be felt through its wrapping
of tissue paper, and which he carried about with
him in his purse together with two fifty-centime
pieces, several stamps, a trunk-key, and a ribbon
of the Iron Cross. The medium made some
striking assertions with reference to these articles
and their history. On closer examination of
the reports of these experiments we find a jumble
of visions and acoustic phenomena (the medium
hears a voice saying something to him), and
finally we get purely intellectual perception, as,
for instance, the declaration " from a strange
country, without a doubt." Acoustic phenomena
may possibly be explained as an alternative expression of conscious knowledge, though this
can surely not be asserted with regard to visual
phenomena. The paramnestic theory is also
applicable to Tischner's case of B, where the
situations remembered were not due to " knowledge," but to sensation of " sight " (not to say of
" hearing "). In this case we must assume that
the original telepathic perceptions themselves
were reproduced, just as we ourselves in some
cases remember events, in others, recall an actual
concrete sensory picture of them.

We conclude, provisionally, that Mrs. Piper's achievements were confined to her intercourse with the living (including those who passed over while her memorizing powers were still unexhausted). It is, however, most desirable that further research should be undertaken to see if this conclusion is correct. Another medium reported on by Osty is said to have visualized prehistoric landscapes and catastrophes on being handed a fossilized animal's tooth, and on touching an antique jewel. This same medium described facts of ancient Greece, though here, of course, verification is extraordinarily difficult. Some one was aware that the objects in question were a fossilized tooth and an antique Greek jewel. And there are enough sources, " conscious " and otherwise, on which a medium of some education can draw for descriptions of geological and historical events. It is only if the visions of the medium exceed these limits and disclose facts which have to be verified afterwards that we should be justified in assuming that psychometry differs specifically from an elaborated telepathy as described above. As I have already indicated in detail, it is, however, possible to ascribe all " historical psychometry " to telepathy. It is only necessary to assume the existence of a subconscious telepathic nexus between all, or at least most, of the medially disposed individuals. In this manner the experiences and knowledge of these people would be inherited from generation to generation, and a perfect medium would thus be able to recount the adventures of Rameses the Great or of Alexander. He might become the spiritual witness of the erection of the Pyramids and of the invocation of Jupiter Ammon. History would thus have direct connexion with the past by reawakening in the souls of men the actual traces of past ages through the intermediary of the great mediums. What a perspective is opened out by the thought that the day may come when the battle of Marathon or the appearance of Socrates before his judges might be described to us by a person in a trance. We should learn everything: how Greek was pronounced, and how Socrates and Plato conversed together; for the voice and physiognomy of the medium of genius is as malleable as wax.

But how would it be if the medium were capable of still greater efforts, and could describe events of the prehistoric age ? If the whole of the past were to be unrolled before us ? The thought is too phantastical, but we are not aware of the bounds of psychometry. The possibility must be recognized and investigated. It is obvious that the truth will take long to establish. If the result of the investigation were to establish the theory as fact, it would mean that psychometry cannot be founded (or at any rate not alone) on a telepathic nexus of humanity. Its causes would be deeper and still more wonderful. Either those would be right who are of opinion that all events leave traces on the object under observation, and that these traces produce corresponding thoughts or manifestations in the psychometric medium, or it must be accepted that these mediums get their telepathic knowledge from the memory of God or that of another superhuman spirit (the Earth Soul of Fechner).

Anna Katherina Emmerich, who was canonized by the Catholic Church, was accredited with supernormal faculties, and in her case what appears to be historical paramnesia has been proved with comparative accuracy. The poet Clemens Brentano has collected a good deal of material about her. She left whole cycles of visions about Jesus and Mary, purporting to contain information on archæological details in Palestine, which were still unknown in her lifetime, but which are said to have been verified lately. Should these assertions really be confirmed—but I confess that I have felt so sceptical about them that I have not even troubled to examine them closer—they would be of the greatest interest for the further development of Parapsychology.

CHAPTER III

CROSS-CORRESPONDENCE

THE store of mediumistic phenomena was further increased some ten years ago by a new development, hitherto unknown, that of Cross-Correspondence. It was discovered by the distinguished secretary of the British Society for Psychical Research, Alice Johnson, who, while studying the automatic writings, of the different mediums, became aware of a strange relationship between them. In some cases this consisted of striking allusions made by one written communication to the other, in the use by both mediums of the same strange expressions, in a common reference to a certain literary quotation, and so on. This relationship was of too frequent and systematic a character to be merely due to chance, and did not necessarily exist between two mediums only, but between several. For instance, on April 8, 1907, Mrs. Piper uttered the words "Light in the West" while in a trance in London. On the same day, three hours later, Mrs. Verrall, a medium in Cambridge, wrote automatically among other things : " Rosy is the East, etc.

You will find that you have written a message for Mr. Piddington, a message that you have not understood, but that he has. Tell him this." Moreover, on the same day, a little later, a third medium, in Calcutta, Mrs. Holland, wrote : " This exceptional sky, beneath which dusk renders the *East* as beautiful and shining as the West, Martha became Mary and Lea Rachel." Closer analysis of these expressions and of their contrast proved that all three scripts were related to each other.

A second instance : On August 6, 1906, Mrs. Holland wrote in India at the end of a fairly long communication, separated by a wider space and in an altered hand :

> " Yelo " (scribbled).
> " Yellowed Ivory."

Two days later Mrs. Verrall wrote in Cambridge on August 8 :

> " I have done it to-night y yellow is the
> written word
> yellow
> yellow
> Say only yellow.'

And her daughter also wrote automatically at the same time, without her mother's knowledge :

" Camomile and resin the prescription is old on yellow paper in a box with a sweet scent."

In other cases automatic writings supplement each other, and only make coherent sense when added together. It is—to use a metaphor— almost as though a manuscript had been cut into scraps and handed to various compositors who would only be able to make sense of the whole after joining the fragments together. Oddly enough, cross-correspondence first showed itself suddenly among a number of mediums, including Mrs. Verrall, Mrs. Holland, Mrs. Piper, and others.

Spiritistic interpretation sees in cross-correspondence the best of all proofs of its teaching that mediumistic phenomena emanate from spirits, arguing that the relationship between the various automatic scripts can only be the outcome of an intelligence beyond the ken of the mediums, which uses the latter to prove its own independent existence through the cross-correspondences. Only an intelligence, it is argued, would be capable of meting out a consecutive idea into distinct parts and then directing the pen of the various mediums so that each should write separate fragments of the whole. Spiritism further points to the strange coincidence that cross-correspondence appeared for the first time after the death of Myers, one of the most eminent scientific English-speaking spiritists, who was expected to furnish a conclusive proof of spiritism. In the first cross-correspondence, the " spirit " purporting to be Myers draws direct attention to the new development and the prospect of its further continuance. As a matter of fact, it is not possible not to regard certain cases of cross-correspondence as evidence of the most remarkable and difficult parapsychic phenomena. It is easy to understand that when confronted with cross-correspondence, scepticism should lose its assurance, and that those spiritistically inclined should become definite converts. It is obvious that cross-correspondence must be attributed to a reflecting mind. There can be no question of chance, for the varied inspirational utterances are too numerous, too striking in character, and fit into each other too well. Despite this, they need not be regarded as any incontrovertible proof of spiritism. The hackneyed contention that the various mediums concerned have come to an understanding with regard to a common deception cannot, of course, be maintained. There is no cause for suspicion here. The possibility, however, is not to be denied that there may be an unconscious telepathic understanding of that kind. We have gradually collected so many proofs of the highly developed intelligence of the subconscious mediumistic psychic life that such a hypothesis cannot be excluded. We know of automatic riddles and of anagrams of such artistic conception that we cannot reject such possibilities. A certain Mr.

A., for instance, while experimenting with automatic writing, at his third attempt to obtain a reply from the supposed spirit to his question : " What is Man ? " received the automatic answer, " Tefi Hasl Esble Lies," of which the solution is " Life is the less able."

It must not be forgotten that the majority of the mediums are confirmed spiritists, so that a tendency or a desire to testify as to the genuine nature of spiritism is ever prevalent. In the same way, in the case of Helene Smith, this tendency was concentrated on the invention of faked languages. Mrs. Piper, however, lacked any such tendency while awake, as her attitude towards spiritism remained neutral. But it must be noted that in her case, the parapsychic manifestations were evolved in trance, in a state of transmuted personality. She was then apparently transformed into other personalities. These " spirits "—i.e. the somnambulistic Mrs. Piper—were, however, as such, naturally convinced as to the truth of spiritism, and their whole activity was concentrated on evolving proofs of their belief. Is it surprising that she was bent on making use of her telepathic faculties to this effect ? Even the fact that the phenomenon of cross-correspondence was manifested with comparative suddenness by the various mediums, is no proof for the spiritistic contention that the spirits agreed to make common use of this new channel. It is a quite sufficient explanation that, once cross-correspondence has been discovered, innumerable mediums should employ it.

No difficulty is encountered in interpreting the cases in which the cross-correspondence confines itself to connexions between automatic script in the way in which a certain word is repeated or referred to. Such similarities must be explained as due to the mediums writing being possessed of telepathic or clairvoyant faculties. It is another matter when one fragment only makes sense when joined to another, each scrap consisting of one sentence. Then it is necessary, unless the connexion between the two scripts is to be regarded in the light of mere coincidence resulting from a purely hypothetical completion of one fragment by another, that a mutual understanding or convention should be assumed to exist between the two mediums to settle which words of the sentence should be written by either. If, however, telepathic possibilities of communication actually exists between them, it is equally admissible to contend that all these various mediums are alike imbued with their desire to add to the proofs in favour of spiritism. It might also be that one medium simply transmits telepathic suggestion to another " à distance," in which case there need be no question of any previous agreement.

A conscious suggestive influence of one trance personality on other individuals would represent a positive novum. *A priori* there is no reason why a person in a somnambulistic state or in a similar condition should not be subjected to suggestion from others, and also subject others thereto. Experimentally, we only know at the present time of suggestion " à distance " (based on the tests of Richet, P. Janet, and others), in the form of suggestions by a conscious individual.

It would be extremely interesting (if it were possible) to persuade the trance personalities themselves to make suggestions either to conscious or to other hypnotized persons. Suggestion on suggestion might also be contrived, by influencing a person under hypnosis, so that he should distribute his own suggestions even at a distance.

Cross-correspondences are from the point of view of logical proof at a disadvantage when compared with other parapsychic phenomena, in so far as we are, in their case, mainly obliged to rely on the veracity of the mediums themselves. Many among them, notably those to whom the most important experiments are due, as also the authors of the reports published in the " Proceedings of the Society for Psychical Research," have themselves supplied the material in full cognisance of the stage reached in the problem at issue. The assumption that the writers must be regarded as common frauds is

5

in contradiction to what is known of their character on the whole; besides which, in Mrs. Piper's case, the phenomena of cross-correspondence were carried out under a system of strict control.

ADDITIONAL REMARKS

In response to several requests I will give below a few more examples of cross-correspondence.

One of the most famous, which occurred right at the beginning of the cross-correspondence, is as follows:

Mrs. Verrall, lecturer in classics at the University of Cambridge, writes:

"On January 31, 1902, I had been lunching with Mr. Piddington in town, and after the arrival of Sir Oliver Lodge from Birmingham was about to walk with them to the S.P.R. Council Meeting at 3 p.m., when I felt suddenly so strong a desire to write that I came down and made an excuse for not accompanying the gentlemen, saying I would drive later. As soon as they had started I wrote automatically in the dining-room the following words:

"Panopticon σφαιρᾶς ἀπιτάλλει συνδέγμα μύστικον τί οὐκ ἰδιδως; volatile ferrum—pro telo impinget."

A few more words were added, when I was interrupted by Mr. Piddington, who had returned, in order to drive with me to the meeting.

All the rest of the day I felt a wish to write, and finally, in the train on the way home to Cambridge, more script was produced. That script contained no verifiable statement but was signed with two crosses, one of them being the Greek cross, definitely stated elsewhere in the script to be the sign of Rector (one of Mrs. Piper's trance personalities). . . .

So far for what happened in England. In Boston, as I subsequently learned, the following took place. At Mrs. Piper's sitting on January 28, 1902, after the reference to my daughter's supposed vision, Dr. Hodgson suggested that the same " control " should try to impress my daughter in the course of the next week with a scene or object. The control assented. Dr. Hodgson said : " Can you try and make Helen see you holding a spear in your hand ? " The control asked : " Why a sphere ? " Dr. Hodgson repeated " spear," and the control accepted the suggestion, and said the experiment should be tried for a week. On February 4, 1902, at the next sitting, and therefore at the very first opportunity, the control claimed to have been successful in making himself visible to Helen Verrall with a " sphear " (so spelt in the trance writing)."

This example is also an instance of the curious and baffling confusion which prevails in much of the automatic writing which contains cross-

correspondences.   Instead of an (actively con-
ditioned telepathic ?) vision which we might
have expected after the séance with Mrs. Piper,
we get at Mrs. Verrall's end (as was so often the
case with her) script mixed up with broken bits
of Latin and Greek (she was a classical scholar),
or, as in the present case, so far as it is published,
script consisting of nothing but bits of Latin
and Greek, in which very clear allusions, obvious
at once, strike us to the séance in Boston ($\sigma\phi\alpha\iota\rho\alpha$
= sphere ; volatile ferrum, telum = spear).

A second example.   On March 11, 1907, at
about eleven o'clock, Mrs. Piper, who was in her
normal waking consciousness, said "Violets.
Dr. Hodgson (said) violets."   In accordance
with previous experience marked utterances of
this kind might be expected to have reference
to a cross-correspondence.   In fact, on the same
day about the same time Mrs. Verrall wrote
automatically :

"With violet buds their heads were crowned.
"Violaceae odores.
"Violet and olive leaf purple and hoary.
"The city of the violet——"

It is hardly necessary to emphasize here the
marked way in which the word violet is stressed.
The whole script seems really to be simply built
up round this word. (This example is taken
from A. Hude's "The Evidence, etc.," p. 283).

To conclude with an example in which several

days elapsed between the cross-correspondence.

On April 8 the Myers' personality speaking through Mrs. Piper, said to Mrs. Sidgwick: "Do you remember Euripides?" "Do you remember Spirit and Angel? I gave both. Nearly all the words I have written to-day refer to messages I am trying to give through Mrs. V——." Mrs. Verrall had already on March 7 done a long piece of automatic writing in which the word "Hercules Furens," and "Euripides" are found.

And on March 25 she had written: "The Hercules play comes in there, and the clue is in the Euripides play if you could see it." Also she wrote on the same day a separate piece of script in which the word "shadow" occurred several times: "Let Piddington know when you get a message about shadow. The shadow of a shade. That is better umbrarum umbras σκιᾶς εἴδωλον was what I wanted to get written." The word "spirit," however, was not used. On April 3 an effort was clearly made to reach a satisfactory conclusion, although the word "Angel" could not be reached. "Flaming swords—wings or feathered wings come in somewhere—Try pinions of desire. The wings of Icarus—Lost Paradise regained—his flame-clad messengers (she draws an angel with wings) that is better F W H M has sent the message through at last."

The cross-correspondence, moreover, was extended to include Mrs. Holland. On April 16 she wrote automatically a passage in which were found these words : " Lucus Margaret To fly to find Euripides Philemon." The names Lucus and Philemon seem to be derived from Browning's translations of Euripides' Hercules Furens. (A. Hude, p. 285).

Many other cross-correspondences, some of them extremely striking ones, cannot be quoted here because of their complexity and of the space which they require for interpretation and comment. The peculiarity of the cross-correspondences from English sources is that they are mostly of an especially contorted kind. Some French cross-correspondence to which I am unable to refer, are (as I am told) much easier to see through. (G. Geley, " Contribution à l'etude des Correspondances Croisées." Documents nouveaux, Paris, 1914).

CHAPTER IV

EUSAPIA PALLADINO—TELEKINESIS

Physical Mediumship

A NEW group of phenomena which is now to be examined, and which is quite distinct from those just discussed, upsets our traditional conceptions still more completely. I refer to the so-called physical manifestations of Mediumship. They also like the phenomena already described, are not found for the first time in the surroundings only of modern occultism. We already find descriptions of such phenomena in ancient literature—for instance, in Josephus. The account of Christ's " walking on the sea " must be included in this category, together with the legends of the middle-ages describing the appearances of persons floating in mid-air wrapt in ecstasy. Similar cases are also to be found amongst primitive men and savages.

It stands to reason that we are much more suspicious of physical phenomena than of the purely psychic. The stability of our conventional scientific conception of the universe appears—possibly erroneously—to be much more seriously jeopardized by them than by new facts of conscious life. Consequently our instinctive opposition to the recognition of abnormal physical phenomena is far stronger than it is to that of psychic supernormal phenomena. Only two of the older mediums connected with such manifestations are still remembered in the present generation, and even so, they are remembered, not so much for the singularity of their phenomena, as for the reason that the phenomena were vouched for as real by scientific investigators of the highest standard.

One of these mediums was Slade, an American. The astro-physicist, C. F. Zoellner, spent much time in experimenting with him, and was helped upon occasion by scientific friends such as Wilhelm Weber and Fechner. Both these scientists declared themselves convinced of the reality of the phenomena as recorded by Zoellner. Unfortunately the plan conceived by the astro-physicist Vogel was never put into execution. He intended to hide in a cupboard and watch Slade closely through a hole in it. On the other hand, we have a report by Dessoirs on sittings with Slade, in which he states that he was fully convinced of the objectivity of the manifestations. These were very varied in character. For instance, Zoellner is stated to have taken two slates, and, after putting a little slate-pencil between them, to have tied them tightly together with string. When Slade then held them under the table in the presence of the sitters, a curious scraping sound was heard. He then moved the package from beneath the table, and after loosening the string, writing was found on the slates. In other similar experiments, it was stated that on a soot-covered surface the imprint of a bare, or only partly, clothed human foot appeared, differing in size to that of Slade's. Moreover, Slade did not always hold the slates himself, and this fact is confirmed by Dessoir.

Other tests consisted, according to Zoellner, in abstracting certain articles from locked, unopened receptacles, or again replacing them there. Then knots were tied in a string of which both ends had first been sealed together, and finally Zoellner describes how two wooden rings, each turned from a single piece of wood, were placed by Slade round the foot of a centre table without unscrewing the top, which was balanced on a column ending in three legs, none of which also were unscrewed. All these assertions are illustrated with photographs, which Zoellner prints in his report.

The objections made against Zoellner's reports from the present-day standpoint are based, in the first place, on the absence of any minutes of the experiments—a most regrettable omission ; and next, on the supposition that sufficient care was not taken to prevent deception on Slade's part. For instance, Slade may have had an opportunity of abstracting the string with seals attached, and substituting for it at the next sitting a second sealed string with knots in it. Again, with regard to the slates : Zoellner has been blamed for not taking sufficient precautions to make an exchange impossible, or to prevent the knots being loosened sufficiently to introduce some sharp, thin object between the slates with which written characters might have been traced. These explanations are not, however, applicable when it becomes a question of emptying sealed boxes; still less so when it comes to placing wooden rings round the legs of a table. It is a significant sign of the weakness of the criticisms against Zoellner, that no reference is made to the facts which are hardest to account for.

I have often made vain attempts to discover whether the table with the ring still exists. A relative of Zoellner, his eminent biographer, Fr. Koerber, my former mathematics master at my " Gymnasium," could only tell me that the table still existed up to a short time ago, but as he could not indicate its present whereabouts it has not been possible for me to ascertain whether the ring actually did only consist of one piece ; also, the exact details of the construction of the table, whether the top was easily removable, etc. In my opinion there is only one possibility of fraud with regard to the table experiment : that Slade hypnotized Zoellner, who was alone with him when the feat was accomplished, unscrewed the top, placed the ring round the foot, and then awakened Zoellner, possibly under the influence of the definite suggestion that the latter should not remember anything that had taken place. Given such conditions he would indeed have had the opportunity of carrying out the most astounding feats. But if this explanation is approved with regard to the table experiment, it would also apply to all the others ; and upon this view none of the tests described by Zoellner, and of which he was the sole observer, can be looked upon as conclusive. The only question is whether such an explanation is acceptable.

To answer this question in the affirmative is made extremely difficult by the facts that, in the first place, the wooden rings were suspended on a sealed cord ; that guests were waiting in the adjoining room, and that, according to Zoellner, the whole proceedings did not last more than five minutes. Consequently, Slade must either have replaced the cord, on which another ring was also suspended, by one exactly similar, or renewed the seals with Zoellner's cipher or a facsimile thereof.

Again, what view are we to take with regard to those cases where there were other witnesses, such as Weber, Fechner, Wach, etc. ? Here we are bound to admit that so far no precedent has been found whereby it has been proved possible to put the company present, consisting of several individuals, into a hypnotic state so easily without their consent, as must (it is necessary to assume) have here been the case. It could only be done by telepathic suggestion. This has been proved to be possible in some instances, but only in connexion with individuals who had already been under the hypnotic influence of the " suggestor," or, in the case where both persons showed signs of parapsychic structure. Upon this view, then, a parapsychic phenomenon at any rate took place. Attempts have also been made to suggest that reputed performances of Indian Fakirs, who, in the sight of many spectators, claim to throw a rope into the air, and make a boy climb up and disappear with it, are the result of mass hallucination produced by telepathic suggestion. It must, however, not be forgotten that this, too, is but a hypothesis, and that a hypothesis is not strengthened by being applied to many cases instead of limited to one.

Zoellner's reports have repeatedly been summarily dismissed on the ground that Slade was— " as all the world knows "—discovered cheating in America. But this rumour still lacks confirmation. Koerber made every effort to clear this matter up, but was unable to procure more definite information. It is, therefore, not justifiable to regard it as an established fact that Slade was a cheat. On the other hand, there is the highly suspicious circumstances that at a sitting with Slade in London, a slate which was supposed to have nothing on it was found to have had letters already written on it, when it was forcibly snatched away from him a moment or so after he had held it under the table.

Unfortunately, Helmholtz refused to examine Slade, despite repeated invitations to do so. His testimony would have been of the greatest value, and the case of Slade would be much the clearer for it to-day. As it is, we only know that Helmholtz's opinion *a priori* was that it was all a fraud, though *a priori* judgments have no significance so far as science and psychology are concerned. The case of Zoellner-Slade must in consequence be left in suspense. But whatever one's opinion may be, the brilliancy and interest of Zoellner's theoretic interpretation of the experimental results, based on the hypothesis of the Fourth Dimension, which he considered proved, remain unimpaired. Its simplicity savours of genius.

The experiments of the physicist Crookes made less of a stir in Germany than in England. He believed he had established the fact that the medium, Home, was able at will to decrease or increase his weight. And in the case of another medium, Florence Cook, he even claimed to have observed and photographed genuine materializations (Katie King). Unfortunately the experiments with Home were not repeated by anyone at the time, despite their great interest, and the comparative ease with which they might have been carried out. Quite recently, a young Berlin engineer, Grunewald, is said to have achieved the same result with another medium, though this has not yet been verified.

Some years after Slade, Home, and Florence Cook had been forgotten, a new medium began to awaken the interest of the European spiritualist world—Eusapia Palladino. Her fame remained undiminished to the day of her death. Dragged through all Europe and half America, surrounded by a galaxy of savants and dillettanti, she has been the theme of a whole literature in every civilized land. And yet, no complete unanimity appears to exist even to-day between the different observers. Opinions are not only divided as to how far the phenomena were genuine, but also as to whether the whole thing was or was not a fraud. All the same, it is only fair to say that all those who observed her for any consecutive length of time are agreed that the major part, and particularly the most striking of her phenomena, were genuine. It is due to this circumstance that the case is of such surpassing interest.

Eusapia Palladino was born in 1854 in a small village in the Abruzzi Mountains—the only child of an inn-keeper. Her mother died at her birth, and when she was eight years old she also lost her father, who was murdered by brigands. Eusapia was put in charge of her grandmother, who brutally ill-treated her. Later on she became a sempstress. After her marriage she gradually relinquished her former occupation for that of a professional medium.

I do not know what her regular income was, but she was repeatedly invited to undertake lengthy journeys to Munich, Paris, London, Petrograd, and she also gave sittings in America. She died in Naples in 1918, a great loss to psychic research. Her medial faculties are said to have developed during her puberty. Between the age of 13-14 she first saw visions, and objects are said to have been moved in her presence without her touching them.

Her temperament—and this is borne out by her portraits—is said to have been joyous, gregarious and inclined to emotionalism. She had no schooling whatever, and could hardly write her name. Nevertheless she was gifted with great natural intelligence, and evinced great knowledge of character in her intercourse with the people with whom she came in contact. In this respect she showed faculties which it is preferable not to find in a medium. To get a sitting with her does not seem to have been too easy. She was fully aware of the part she played in the world. " E *una* Palladino," she was wont to say of herself, and she insisted on being treated as a great lady in spite of her want of culture. She did not always submit to the conditions of control to which attempts were made to subject her, but sometimes autocratically imposed her will with regard to the manner in which she wished the sitting to be held. For this reason, sittings with her were more often confined to mere observations than devoted to actual experiments. For, whenever pressure was put upon her or she was contradicted—however substantial or well-considered the reason might be—the investigator took the risk of having the sitting abruptly broken off, and the case was of such interest, and as a rule so much expense had been incurred in obtaining the phenomena at all, that it was usually preferable to be content with mere observation, as soon as Eusapia began to remonstrate. The impression left by her own high opinion of herself was counterbalanced by her kind-heartedness. Her own early fate, at the remembrance of which she often shed tears, made her charitable, particularly to orphans. Having been given the choice of a present, at a sitting one day, she begged for an artificial limb for a child whose own was about to be amputated. She avoided solitude, and loved to have company around her always, as she was often made uneasy by her own phenomena. She was in actual fear of darkness, and always kept a night-light burning;

she even preferred not to have all the lights extinguished during the sittings.

Though other mediums, of whom there are similar reports, such as Mdme. d'Esperance, Frau Pribytkoff, etc., never, or but rarely, emerged from the spiritistic sphere in which they had been discovered and which they looked upon as their spiritual home, Eusapia Palladino was examined by a considerable number of scientists in whose rank, Germans were but sparsely represented.

The examinations carried out in 1905–08 in Paris at the Institut Général Psychologique were distinguished by the presence of the great number of eminent investigators who took part in them. The report published by Courtier, Professor of Psychology at the Sorbonne, repeatedly mentions the names of Perrin, Poincaré, Curie, Bergson. The investigations (there were forty-three sittings in all) were conducted in the manner usual, not only with Eusapia, but also with other psychical mediums. A corner of the room was partitioned off from the rest by a black curtain fixed with metal rings to a pole. Eusapia sat just in front of the centre of the curtain behind a table, with a sitter on either side of her who had instructions to watch her hands and feet. They were each told to hold her hand—preferably by the thumb—and to place their left on her right

6

foot and their right on her left foot respectively. Unfortunately, it was she who often placed her own foot on that of the observer. Then, a so-called " chain " was formed by all the participants of the séance by joining hands right and left round the table. Eusapia refused to be bound, no matter how lightly, as she declared that this reminded her of a lunatic asylum, and gave her the feeling of being mentally afflicted and forcibly tied down. Neither would she permit any flash-light photographs, though as a matter of fact the lights were said to have been so bright at times, that it would have been possible to read by them. In the interior of the cabinet, behind the curtain, was a light table for bric-à-brac, some smoke-blackened articles or papers, a jar filled with modelling clay or putty, and a zither.

The sittings usually started in a bright light which gradually was made fainter. According to the minutes, the manifestations began with various sounds of unknown origin on the table—raps as if by a finger, scraping as by a nail, etc. As the lights grew fainter, the objects in Eusapia's vicinity began to move about spontaneously. The table rose (" levitations ") and the various objects in the cabinet were heard to change place. With a still dimmer light, it is asserted that vague outlines of hands and other parts of the human body, such as a head and bust, became visible near Eusapia, appearing through the central and side folds of the curtain. Brilliant dots or sparks resembling electricity were sometimes seen. A phenomenon which often recurred was that the curtain behind and next to Eusapia billowed out, and to the touch felt as if it was pushed forward by something tangible. Phenomena of other description were also reported by the observers. Equally remarkable were the imprints left upon the modelling clay, for instance of human veil-covered hands, greatly resembling those of Eusapia, or occasionally of a human face. Part of the records were obtained by a registering apparatus, but the most important are based on the evidence of the sitters. The report of the result of the investigations is summarized in the ten following points :

(1) Displacements (backwards or forwards) and the (complete or partial) levitation of certain heavy objects (ordinary small tables) in Eusapia's vicinity were evidenced by registering apparatus.

(2) Some of the said movements of objects appear to have been caused by the mere touch of the hands or clothes of the medium, and even without her touching them at all. During the complete levitation of the table before which she sat, or of the smaller table placed near her, her muscles were strongly contracted. But she did not seem to will to elevate the objects in the same way that we ordinarily will things.

(3) The supporting point of the force which

raised the objects seemed to be centred in the medium, as the scales on which she was placed during the elevations marked an increase and decrease in weight which corresponded to the laws of mechanics.

(4) It was shown that she could discharge electroscopes from a distance.

(5) It was shown that she could cause molecular oscillations from a distance (rapping, sound-vibrations).

(6) Lights of unexplained origin were seen near the medium during the sittings. Some of these phenomena were like electric sparks.

(7) Those present said that they observed human forms and felt themselves touched. But it must also be noted that fraud has been proved with regard to some manifestations of this kind.

(8) In the course of some of the sittings, Eusapia passed into a secondary condition of an unstable type. She complained of hyperæsthesia to the touch during the greater part of the sittings, and for some time after them also. She complained further of partial amnesia with regard to the séance phenomena.

(9) The ideas and the will of Eusapia influenced the nature and the course of the phenomena.

(10) Fraud was possible, but to what extent it was practised is hard to determine.

This brief synopsis can certainly not take the place of the detailed report in the original. But it proves conclusively that the above-mentioned eminent investigators, who had Eusapia again and again under their observations were convinced of the objectivity of part of the phenomena. Their testimony is naturally of far greater value than the judgment of some obscure minor scientist of Frankfort or elsewhere, who delivers judgment on Eusapia without personal experience.

On the other hand, according to Courtier's report, any conclusions about Eusapia are necessarily to some extent vitiated by the certainty that some of her manipulations were fraudulent. On one occasion, shortly before a sitting, she was seen tampering with a pair of scales, which she was manipulating with the aid of a white hair. On another, a little nail fell to the ground, apparently from Eusapia's left hand, who evinced great surprise. The nail could be used to make marks on blackened paper similar to those found at the sittings. In the total darkness of one of the séances, while the sitters felt various touches, Eusapia freed her hand with lightning speed from that of Courtier, and immediately after, as Courtier recovered from his surprise, Eusapia's hand lay once again in his. In short, there is not the slightest doubt that she practised fraud, a fact of which other investigators were also convinced. At the first investigations undertaken by the British Society for Psychical Research, who make a rule of never continuing experiments with a medium convicted of fraud, the investigations with Eusapia were abandoned on that ground. The only surprising part consists in the repeated assertions of investigators that Eusapia's deception was of so childish a nature that it could not be taken seriously, when we have regard to her intelligence, and remember that phenomena often occurred at the same time which could not possibly be due to fraud. True, such phenomena always took place in Eusapia's immediate vicinity. Once during the Paris séances, when a flashlight photograph was taken of her against her will, it showed her with an extremely crafty expression. And yet these incriminating circumstances are again outweighed by others of such importance that scepticism must of necessity cease. Thus the Paris report : " Eusapia made a movement of her hand, and the zither sounded from within the cabinet. Eusapia scratched the hand of M. d'Arsonval with hers, and again the zither was heard, as though plucked by fingers." " Another time a small board which had been nailed to an inner corner of cabinet was torn from its foundation." Many far heavier objects were moved, lifted and transported, for instance : a stool was raised one metre high, and a dish full of putty placed on top of it. This stool stood in the cabinet, and through a gap in the curtain it was seen to advance and retreat several times. The wish was expressed that the dish with putty should be lifted on to the table. Eusapia requested that all should concentrate their will on this idea and it would be realized, and realized it was. The stool was then hoisted on to Monsieur Curie's shoulder. The receptacle with the putty weighed seven kilos, and it took considerable strength to lift and hold with one hand. The dish was 30 cm. long and 24 cm. wide (Controllers—left : Mr. Komyakoff ; right : Mr. Curie)." " She was able to depress a letter-weight in full light without touching it ; but when the scales was placed under glass so that Eusapia could not possibly touch them with a thread, they did not move.   On the other hand, the balance was again depressed after the glass cover had been removed under conditions of observation which certainly seemed quite adequate.   Her hands lay to the right and left of the scales ; on lowering them the scales sank in sympathy."

We ask in vain how such phenomena could have been achieved by fraud, without detection by the sitters.   This explains how it was that the Society of Psychical Research made an exception in the case of Eusapia, and arranged for a renewed examination of the medium through several of its most experienced members in Naples. These investigators were convinced of the genuineness of the phenomena, Carrington among them, who was so well known for having shewn up innumerable pseudo-mediums.

I exclude the reports of Lombroso and Flammarion from the others. The latter may be called a Visionary; and in many of the works of Lombroso, particularly in his book on Genius, inaccuracy and superficiality are so conspicuous, that he must be taken with a certain reserve. But there are still the experiments made by Botazzi in Naples, who was the Professor of Physiology at that University, and who collaborated with five other professors at the university and polytechnic. According to this report, the reality of the phenomena was definitely proved under good conditions; for instance : " Both of us, Mr. Scarssa (Lecturer of Physics at Naples University) and I, kept our eyes fixed on the mandoline, and we can definitely assert that the instrument, clearly illuminated by the lamp above it, was not touched by the visible hands of Eusapia. The latter was sixty cms. away from it, but the mandoline moved as though set in motion by magic. It is impossible to describe the impression made by the sight of an inanimate object moving in dead silence, not only for a second, but for several minutes, without being touched by anyone, under the compulsion of a mysterious force, among other inanimate objects."

On another occasion, Botazzi (who weighed eighty-nine kilos) was propelled along the ground with the chair on which he was seated. During a sitting in Munich, at Schrenck-Notzing's house, the table before which Eusapia sat was elevated, while her right hand was controlled by Professor G., and her left by Dr. Albrecht, while Schrenck-Notzing lay under the table in order to keep her legs and feet under observation. Another time there was even a lamp under the table. At a séance at Rome, while the hands of the medium were controlled by the physiologist Professor Luciani and the alienist Sante de Santis, the curtain of the cabinet behind Eusapia was inflated some twenty times in succession. It was possible to touch the curtain, lift it up and also put one's hand between Eusapia and the curtain. And when, in the course of independent experiments connected with active telekinesis the light was suddenly switched on, Eusapia was discovered to be in a deep trance, her hands held by her neighbours.

Flournoy reports further : " It is to Richet that I am indebted for the privilege of having taken part in several séances with Eusapia Palladino last year (1898). The conditions of control then were such that there is no room for doubt, unless we are to distrust the combined testimony of sight, hearing and touch, as well as that modicum of critical sense and astuteness, in the possession of which every person of ordinary intelligence prides himself. The only other alternative is to assume that there were secret doors in the walls of Richet's work-room, and that he, together with his learned assistants, were the wicked aiders and abettors in the farce enacted by this charming Neapolitan lady."

According to the material at my disposal, the following university professors, among others, made experiments with her : the physiologists Richet (Paris), Luciani (Rome), Botazzi (Naples), the alienists and neurologists Sante de Santis (Rome), Morselli (Genoa), Lombroso (Turin), the anatomist Pio Foa (Turin), the scientists M. and Mdme. Curie, Perrin, Poincaré (Paris), the astronomers Schiaparelli (Milan), Flammarion (Paris), the psychologists and philosophers Courtier and Bergson (Paris), Flournoy (Geneva). All these investigators are convinced of the genuine nature of certain supernormal phenomena as demonstrated by Eusapia Palladino. Is there really any sense from a scientific point of view in those who have not been observers persisting in face of this evidence in considering the non-existence of the phenomena in question as more probable than their objectivity ?

Only those investigators who casually attended only one or two sittings as, for instance, Dessoir, Lipps, Munsterberg, Moll, were still sceptical. But their evidence—that of Dessoir is very

shaky—is of little importance, as their investigations were very slight in comparison with those
of investigators who were able to follow the case
at length and in detail.

It should also be noted that Eusapia's
mediumistic faculties were obviously variable.
The same English investigators, who, in 1908,
came to positive conclusions, were present at
various quite negative sittings in 1910 in Naples,
in which Eusapia did nothing but cheat. This
circumstance really is in favour of the justice and
objectivity of their first report.

Those who put the whole thing down to fraud
support their case by the outcome of two
American séances. At one of them, Munsterberg arranged that some one should crawl along
the floor towards Eusapia without her knowledge,
and seize hold of her suddenly in the darkness
during the séance. This resulted in an ear-
piercing scream from Eusapia and the abrupt
breaking-off of the sitting. The person under
the table declared that he had seized " an unshod
foot." When the light was turned on, Eusapia
was seen to be fully clothed. During another
sitting with Professor Lord at Columbia University, two observers claim to have noticed how the
objects in the cabinet were set in motion by
Eusapia herself, who had managed to free one
foot from control.

Important though these assertions may be,

they do not explain the Paris observations, as well as many others, and the uncertainty is increased by the divergent opinions of the conjurers who were consulted. An English conjurer declared that Eusapia's manifestations were absolutely genuine, and that certainly could not be reproduced by conjuring. On the other hand, two of his American colleagues pronounced very unfavourably against them, and insisted that her performances were all faked. The point argued by Eusapia's partisans is that, like all mediums, she was greatly irritated by the ostentatious display of mistrust, which caused a considerable diminution of her psychic faculties, and it is also emphasized that she was never equally consistent in her performances.

Another American investigation undertaken at Columbia University can also be set against the arguments of Munsterberg and Lord. An onlooker was able to watch the whole time through a hole in the top of the cabinet, and so discover whether Eusapia moved the objects contained therein by means of a hook or other fraudulent contrivances. His report is that at every sitting a new organic member—a pseudopodium—appeared from under the curtain behind the back of Eusapia with the aid of which apparently the mechanical effects were produced. But similar reports also had already been published during the 'nineties concerning the pseudopodia which apparently emanated from Eusapia's body. Consequently those who are opposed to the theory of fraud insist that in the case of the Munsterberg exposure, the foot was not Eusapia's own, but a pseudopodium.

These observations which until lately seemed highly problematical, and more or less a subterfuge, have received further support in the conclusions arrived at quite recently by the English physicist, Crawford, by Ochorowicz, late professor of philosophy in Warsaw, and by Schrenck-Notzing. Crawford declares that in the case of an Irish medium, he has repeatedly proved the presence of certain " rod-like " projections of varied lengths and thicknesses which, though invisible, were perceptible to the touch and felt cold, sticky, and like reptiles. Schrenck-Notzing and Ochorowicz were able, in the case of another medium, to photograph similar projections on some occasions, though on others they remained invisible. Thus, another explanation of the cases where Eusapia was charged with the fraudulent manipulation of what looked like threads, might lie in the assumption that these were actually composed of fine organic rays.

The telekinetic movement of objects, the levitations, as well as the strange touches experienced by the sitters, particularly in the dark, are said to be produced by these pseudopodia ;

they are able to become quite rigid, and by a purely mechanical process fasten themselves on to the objects moved. During levitations accordingly the weight of the medium is invariably increased by that of the objects raised by the pseudopodia, as has been repeatedly established.

Upon this hypothesis there is no question of the direct working of the mind of the medium upon distant things, and if the working of her mind is regarded as confined to her own organism, physical mediumistic phenomena are easier than formerly to fit into our normal conception of the universe. Distant objects may be treated as set in motion by means of pseudopodia, which themselves behave in a quite normal mechanical manner. The problems of telekinesis and levitation is thus relegated almost entirely to the organic sphere, in so far as the so-called pseudopodia are of organic nature. The observations of the above-mentioned investigators are, when taken in conjunction with those of older date, of such a momentous character that it becomes imperative to verify them still further objectively. We are possibly confronted by an entirely new category of psycho-physical phenomena, which makes the dependence of material or semi-material events upon the action of the mind far greater than was ever dreamed of in the past. Though even so, I should like to remark that the theory of pseudopodia does not in all cases suffice.

It is noteworthy that the medium possesses the sensation of touch through the pseudopodia. It is from them that Eusapia derived her power of perception. Innumerable references in the reports of Botazzi and others indicate that she knew when she moved an object at a distance or made an impression on clay. It is therefore no proof that she was cheating because she emitted a piercing scream when her foot or the pseudo-podium's was seized ; it is equally possible that she actually did feel the pain of the vigorous grasp on the foot and Crawford with his medium claims to have shown that this is true.

It is still quite uncertain of what substance the curious efflorescences are composed and how they are actually formed. The main difficulty con-sists in the fact that the pseudopodia are able partly to penetrate clothing. An analogy to their power of becoming stiff might be found in the sexual organs of mammalia.

But when all is said, the mind of the medium is still apparently the deciding factor—the sole means through which all the strange phenomena of physical mediumship are evolved. In some cases this fact stands clearly out and Eusapia herself was conscious of it. In such cases she could predict what was going to happen. For instance she made a gesture of striking or twanging strings, and the sound of rapping or of the mandoline was heard. Botazzi paid particular attention to these connexions, though reference thereto is also found elsewhere. It must remain an open question whether conscious mental processes in all cases so far as Eusapia was concerned preceded the phenomena, but it is obvious that normal ideas, thoughts, and acts of will, etc., cannot alone produce these remarkable effects. There must be also other conditions still unknown to us which have to be fulfilled. To invoke the " subconsciousness " of Eusapia as the deciding factor, is to make use of an entirely insufficient conception. Subconscious processes have no more influence on material things than conscious ones, and if these other unknown conditions were unfulfilled, the subconscious processes would have no more effect in Eusapia's case than with other people. The real motive for coming back to the explanation of the " subconscious " lies in the assumption that the actual " vital factors " which differentiate organisms from purely physico-chemical formations, are not to be found in special independent faculties of any kind, but are to be identified with the unconscious acts of the soul itself. If, therefore, there are unconscious mental processes which build up the organism by their influence upon inorganic processes, the inference is that the efflorescences, pseudopodia, etc., are constructed

by similar, or at any rate, similarly unconscious
acts of the mind. This assumption may be
correct, yet these temporary limbs, as well as
our ordinary members, appear to be nothing but
the tools used by the ego in its conscious acts.
The fact that Eusapia was able to determine in
advance the nature of the phenomena and rap
out a given number of knocks on the table in the
cabinet in accordance with the wish of a sitter,
does not alter the explanation. It is of no matter
whether we are concerned with genuine manifestations of will power or not. Eusapia denied
it and we cannot disprove her denial. As a
matter of fact, however, there are many happenings which seem to be the result of the will and
yet are not so. For instance some people perspire
with supernormal facility; they are able by
concentrating their minds on perspiration to
produce drops of sweat on the palm of their
hand, and yet this cannot properly be called the
result of an act of will. The solution of the
physical phenomena of mediumship will possibly
be found in studying such cases.

The outward form in which the psycho-
physical phenomena were produced by Eusapia
was spiritistic. She herself was convinced of
the truth of spiritism and ascribed the phenomena
to a certain spirit, " John King." In certain
cases he " controlled " her and spoke through her
in a voice somewhat altered from her usual one.

7

Eusapia also upon occasions produced inspirational phenomena. As, however, all the attention of the investigators was invariably focussed on the physical phenomena, the reports regarding Eusapia's psychic condition are unfortunately rare, and I have not been able to obtain an accurate account of the changes which took place in her personal condition. As she was very uneducated and certainly had no original tendency to introspective self-analysis, even were she still alive, there would be but small hope of getting more precise particulars of how she perceived and felt things from her own (subjective) point of view. And even so, whatever might have been possible, her character was not such that we should have been able to rely very much upon her word.

CHAPTER V

EVA C.
PROCESSES OF MATERIALIZATION

THE phenomena of telekinesis mostly found in the case of Eusapia Palladino are not the most striking manifestations of which physical mediums are capable. Another group, which even two years ago seemed to me quite incredible, consists in *Processes of Materialization*. In this group are comprised those cases where, in the presence of a medium, formations of material or semi-material nature are produced in the shape of organic or semi-organic structures of a supernormal kind. Such phenomena were numerous in the case of Eusapia Palladino also.

The problem of mediumistic materializations has again come to the front, by reason of the publication of Schrenck-Notzing's book on "teleplastic" materialization processes in the case of a French medium, Eva C. A parallel, and almost as voluminous a publication was issued simultaneously by a French lady, Mme. Alexandre Bisson, in whose house the medium lives, and who has apparently constituted herself the latter's psychological impresario, inasmuch as the organization and supervision of Eva C.'s trances remain nearly always entirely in her hands.

The phenomena of Eva C., according to the reports of the two authors—whom for the moment we take as our authority—consist in a quasi-organic substance extruded from her organism—generally through the mouth—capable of independent movements in the form of strange shapes. These shapes sometimes resemble parts of the human body, though more frequently they take the form of human faces or figures enveloped in floating veils. The shapes are then said to dissolve before the eyes of the spectator . . . that is to say, to re-enter the organism of the medium.

The flat surface of these formations is their oddest characteristic. The hand which purports to be materialized looks like a fake ; the faces or figures seem to be cut out of paper and subsequently veiled. Sometimes it looks as though there were actual folds and wrinkles in the paper itself. I found this impression confirmed by my examination of the stereoscopic prints kindly forwarded to me by Schrenck-Notzing. Nobody who looks at these pictures without further explanation could think that they are anything else than drawings on paper or material, possibly sketches from illustrated papers brought with her by Eva C. ; yet this hypothesis is denied both by

Schrenck-Notzing and Mme. Bisson. According to the former, Eva C., at each séance was clad in tights which he had bought himself and into which she was sewn ; furthermore, she was carefully searched each time from head to foot, to preclude the possibility of any objects being smuggled in—hair, ears, mouth—in fact, her whole body was examined. She was also submitted to a gynæcological examination. Sometimes she was even entirely unclothed during the séance. The cabinet, too, was overhauled before and after the sittings. In addition to this, very important phenomena and spontaneous movements of the " teleplastic " substance and of the materializations have been observed with the curtains open.

Two doctors—Gulat-Wellenberg and Mathilde von Kemnitz—set up the " Rumination Hypothesis." According to this, Eva C. belonged to those rare people who are not only able to swallow objects whole, but also to bring them up again when they want to do so. Thus, she might have smuggled portraits painted on muslin into her œsophagus, and brought them up again during the séance, when her head was behind the curtain of the cabinet. Schrenck-Notzing's reply to, and criticism of, this theory, has so completely demolished it, that there is no real necessity for further discussion of it. He even examined the contents of the medium's

stomach one day, and it is impossible to explain
how Eva C., whose hands—according to the
reports—were continuously under control, could
have been able to unfold the tightly packed
scraps of pictures which she is supposed to have
brought up again, and finally drape them in
veils. On the other hand, it would not be
right to assert that Eva C. had no interest to
tempt her to fraud, for this would not be in
accordance with the real facts. For years she
has lived in easy surroundings, as a guest in Mme.
Bisson's house, where she feels herself to be the
centre of interest to an ever-increasing scientific
circle. This applies with equal force to Mme.
Bisson, who has also been suspected by some
people of being engaged in fraud. She it is who
is usually responsible for the putting of Eva C.
into a state of trance, and who generally controls
and influences the medium. As Mme. Bisson's
voluminous book on Eva C. was published
simultaneously with that of Schrenck-Notzing,
literary ambition and desire for notoriety may
well be the explanation of her participating in a
fraud. The only question is whether her
character is such as to make it likely. Schrenck-
Notzing, says it is not ; and mere suspicion is
no proof. On the other hand, it may be that
Schrenck-Notzing too, cannot be considered
quite impartial. It is much more important
to note that séances held in the absence of Mme.

Bisson have been productive of positive results. It is all the more amazing though, that Mme. Bisson should not have insisted of her own accord on being subjected to the most searching investigations—once suspicion against her was openly voiced—though it is only fair to say that she has permitted a limited examination to be made of her person.

So far, it is clear that the pros and cons are evenly balanced. The impression given by the innumerable photographs taken by Schrenck-Notzing and Mme. Bisson make us incline towards the theory of fraud, though on perusing Schrenck-Notzing's reports the scale tilts the other way. A definite decision is only possible if in the first place we make up our minds whether Eva C. was adequately controlled both before and during the séance, and secondly, whether Mme. Bisson had no opportunity for fraud, and was unable to pass on objects to Eva C. It is of lesser importance for the moment, to ascertain how much reliability can be placed in the reports and minutes of the sittings, as in this case the photographs supply all requisite information. On the other hand these photographs cannot be taken as proofs of the objectivity of the phenomena, as they are only momentary reproductions, and there is no film which reproduces simultaneously the ejection of the substance from the body of Eva C., together with its spontaneous movement and the further phenomena. It must, however, be added that even if the physical control exercised over Eva C. were sufficiently strict, and Mme. Bisson is not a cheat, never was a medium subjected to so many precautions as was Eva C. by Schrenck-Notzing. A further point of great importance lies in the use of a number of cameras at the same time from different angles, and in the taking of stereoscopic prints.

It is, of course, impossible to prove the objectivity of the minutes of the sittings ; their importance depends on our opinion of Schrenck-Notzing. The fragmentary character of the notes is sometimes regrettable, but a change in this respect is hardly possible, in view of the suddenness and rapidity with which phenomena occur.

As I have already mentioned, the most pronounced characteristic of the actual phenomena, consists in their peculiar flatness ; but it is also noteworthy how clearly they are dependent on the mind of the medium. It is as though these materializations were pictures of the imagination or of the memory, for in some cases they are strikingly like published photographs, and in one case even a few letters from the pages of a periodical were reproduced. This circumstance which so strongly favours the assumption of fraud might, if fraud is not at work, almost be regarded as proof that these materializations are no more

than some physically objective transformation of
Eva C.'s memory pictures.[1]

Are there any other circumstances which
might definitely influence judgment, either for
or against ?

As a matter of fact, these do exist as the result
of the new investigations to which Eva C. has
been subjected. She has since the first series of
experiments been examined for a whole year in
Paris, by a psychologist, Dr. Geley, at the rate
of two sittings a week, the sittings taking place
for three months in his own laboratory. As he
was given the opportunity of lecturing on the
subject in January, 1918, in the College de
France, it is clear that in the philosophical and
scientific circles of Paris he must be looked
upon as a serious and reliable investigator.

His observations sweepingly confirm the
striking reports of Schrenck-Notzing. It even
appears that the phenomena concerned had still
further developed and were more easily observed.
In his opinion fraud is not only highly improbable,
but actually impossible, owing to the stringent
conditions of supervision. " I do not say : ' No
fraud took place in these sittings, but there was
no possible chance of its perpetration.' I cannot
repeat this often enough : the materializations
were invariably formed before my eyes ; I have

[1] Another possibility is that these pictures are conveyed to her
telepathically from another source.

observed their origin and development with my own eyes." And more than a hundred other men of science have had the opportunity of witnessing the same phenomena as well. Prominent among them are the names of Richet, Courtier, and Clarapede.

It is of equal interest that Eva C. was examined by a Committee of the Society for Psychical Research in the spring of 1920. Not less than forty sittings took place. Of these a considerable number produced results. The conditions imposed and the measures taken to control the examination were very strict. The conclusion of the committee was that upon the basis of their own observations they were unable to reach an absolutely positive decision ; but that if the earlier observations of Schrenck-Notzing, Mme. Bisson and Geley were brought into account, the verdict must be in favour of the genuineness of the phenomena. The phenomena observed were mostly similar to those previously observed by Schrenck-Notzing and Mme. Bisson, only smaller. This is shown by the photographs. Whilst the photographs of Schrenck-Notzing and Mme. Bisson often show heads as big as life, and sometimes forms upon a scale larger than life, those of the committee are in all cases of quite small objects, apparently of the size of a few centimetres only. In these circumstances the committee was unable to come to a conclusion recognizing the phenomena as genuine without any reservations ; at least, the committee as a committee, was unable to do so unanimously. But it is clear, as is often the case with committees, that the final conclusion is a compromise. The final summing up especially is anything but consistent. It is balanced, very clearly, between recognition and non-recognition. The report has no reservations in its statement that there was upon no occasion any suspicion or trace of fraud. On the contrary, Eva C. made the control easy in every way, and never made difficulties. The control of the medium during the sittings was so good that it was completely impossible for her to have smuggled objects in with her to the sittings and then manipulated them with her hands. If fraud is still admitted to be possible, then the only possible explanation must lie in the regurgitation hypothesis. The phenomena were not big enough for the committee to be able to rule this hypothesis out altogether as impossible. Upon one occasion the " substance " coming out of the mouth of the medium forced its way through a veil which covered the head of the medium, and took a definite shape outside it. This shape was then dissolved, and the " substance " retreated through the veil into the medium's mouth. It would, therefore, be necessary to assume—upon the hypothesis of regurgitation—that the medium had got possession of a wax-like substance which could easily be melted and become hard again. Geley, in his acute criticism of the report, has rightly pointed out that it is quite impossible to believe that no trace of such a wax-like substance would have been left upon the veil. It is not, therefore, surprising to hear that Dingwell, when recently at Munich for the purpose of experiments with Schrenck-Notzing's medium, Willy Sch., declared that he only put forward the hypothesis of the existence of such a substance to make clear the complete absurdity of the whole explanation. This may remind us of Galileo's conduct when, considering that the time was not yet ripe for openly supporting the views of Copernicus, he composed a dialogue between supporters of the old and new point of view in which he allowed the supporters of the old view to prevail by the use of such contradictory arguments that every reader was inevitably convinced of the rightness of the newer Copernican conceptions.

Quite a number of the phenomena and the circumstances under which the sittings took place, make it very difficult to explain why the committee refused to come to a positive conclusion in favour of genuineness, except by an instinctive disinclination on the part of the majority to allow anything to be accepted as proof, except perfectly obvious brute facts upon an overwhelming scale.

This remainder of prejudice acted, as is well known, like a heavy drag upon the sensitiveness of the medium and was, with other disturbing factors, the reason why Eva C.'s phenomena in London were markedly not so strong as in France or at Munich.

Considered by themselves the very careful reports of the individual sittings (when regard is had to the conditions of control and of the unsatisfactory nature of the regurgitation hypothesis) furnish a further proof of the genuineness of the phenomena and are in themselves scientific material of especial importance. The diminutive size of the phenomena, which the committee quite naturally did not desire, itself raises a whole host of interesting theoretical considerations as to the nature of the processes which are in play.

These surprising new discoveries with regard to materialization-mediums affect the question of the connexions between materializations and the pseudopodia by which telekinesis is occasioned. Schrenck-Notzing assumes that they are different stages of one and the same process. This begins with the radiation of the finest thread-like or even shapeless effluvia from the organism. The next step consists in their fusion into more solid formations, and the following one—in their transformation into the flat, sketch-like forms, which have been photographed in considerable number in the publications concerning Eva C. With further progress, the materialization develops into plastic forms, which at first sight are indistinguishable from normal organisms. The materializations are usually invisible in their first stage, although " palpable " ; but it is impossible to lay down any hard and fast rules, as even effluvia of the lowest grade have been photographed. The formations are mostly unable to stand exposure to light, though there are exceptions.

Geley has set up the hypothesis that the substance emanating from the materialization-mediums is connected with the yet undifferentiated fundamental organic matter, and that it takes on shape under the eyes of the spectators, and appears to them as a genuine head, a real hand, etc. He compared this process to the metamorphosis which takes place in the cocoon of the chrysalis or a caterpillar when it turns itself into an almost homogenous primitive organic substance, in order to build itself up again into a new formation—the butterfly. Even were we to admit the existence of such an undifferentiated primitive organic substance which can be fashioned by vital factors, it would only represent one side of the discovery, the other side of which would be the power of the vital force of the medium herself to change its own direction. These changes clearly take place in the medium under the influence of intellectual processes. The construction of hands and feet is evolved according to their conception. On the other hand, it cannot be a question of direct influence of the medium's imagination on organic matter, for the medium has no knowledge of the finer microscopic and ultra-microscopic structure of a hand or a foot. Her conceptions we must assume work through the medium of the vital factors.

That the human mind has some power over vital forces, has, apart from the arguments from materialization processes, been undoubtedly proved. It is well known that it is possible in the case of some people to produce blisters, hæmorrhage, and other phenomena through hypnotic suggestion. However trifling these experiments may appear, great metaphysical significance is to be ascribed to them inasmuch as they afford proof of the action of mental processes on the vital forces. These physiological results would not be possible, if the hypnotic expectation of the formation of a blister or hæmorrhage did not effect a corresponding alteration in the tissues involved. The construction of the organism, however, is the work of the vital forces, and the production of a blister or hæmorrhage is not possible without interfering with the work of the atoms which compose the organism. This interference can hardly be regarded otherwise than as a procedure to which the vital factors so adapt themselves that they rearrange the molecules in the way in which they are arranged in a burn, or in the loosening of the cells which causes the blood to flow. The mere idea of a blister or of a hæmorrhage can hardly be regarded as sufficient by itself to produce either of them, for the idea in itself contains nothing but a representation in thought of a visual image. We must, therefore, interpose factors between the idea and the physical effects —I mean those vital factors which regulate the various processes of the organism. In the cases of the blister and the hæmorrhage we are not concerned with the production of results, which cannot be brought about by normal means. Blisters are ordinarily produced by *outward* burns, though hæmorrhage of course might take place in certain illnesses without outward in-fluence. A better illustration perhaps may be found in cases where under the influence of auto-suggestion, pseudo-pregnancy is occasioned with the attendant changes in the organism, which normally are only produced by the vital factors after conception.

The real mystery of the influence of thought and imagination on the organic material does not only lie in the fact that the mind produces effects on physical things, but also in the fact that milliards of atoms are immediately dis-placed in a completely orderly manner while the individual himself is totally unaware of all the movements which are necessary to obtain the result arrived at. It is just as though we imagine that an entirely ignorant monarch gives instructions for the execution of some great undertaking, and has no idea how it will be carried out. Suddenly engineers, architects, technical advisers, mechanics and workmen start up to begin and complete the work. In like fashion, neither the person under hypnosis, nor the waking auto-suggestor, nor the materialization-medium know how their ideas are actually being put into execution. Despite this, the acting vital forces immediately set intelligently to work in order to achieve the result.

The recent developments have thrown a certain light on earlier reports about mediums, who had not been so thoroughly tested or who had not been tested under scientific control. I am thinking above all of Crookes' medium, Florence Cook, Mme. d'Esperance, and several others. Mme. d'Esperance is the more interesting by reason of the detailed autobiography left by her —probably the only autobiography of a materialization-medium. To be sure, she, too, takes her stand on spiritism. The resemblance in type between all these cases is obvious, despite the individual peculiarities of each of them. Of course, this does not tend towards a strict proof of their objectivity, though the probability is

8

enhanced thereby. If we accepted this probability, we should find in one chapter of Mme.
d'Esperance's autobiography a detailed description of a sudden abrupt interruption of
her state of trance during a complete "act of
materialization." The ostensible reason appears
to have been the wish to catch her cheating,
while according to her own account, a case of
"*dédoublement de personnalité*"—which went
further than in any other known case, and which
actually brought about the biological splitting of
her body—was taking place. Remarkably enough,
she was conscious of her own individuality
simultaneously with that of the supposed materialized spirit, just as Helene Smith was in her
semi-somnambulism, and this might be looked
upon as an extraordinarily strong proof for the
anti-spiritistic interpretation of the whole process
of materialization. The interpretation of this
case given in my "Phenominology of the Ego,"
should be consequently somewhat altered. Unfortunately I do not for the moment possess
Mme. d'Esperance's book, which would enable
me to go into the matter further.

The mind of Mme. d'Esperance "animated"
her own body as well as that of the "materialized" spirit. The explanation resulting from
Botazzi's observations of Eusapia, i.e., that a
materialization-medium feels through its pseudopodia, can thus be extended to the forms material-

ized. Their psychic life belongs in reality to the
mind of the medium. Unfortunately, the sudden
interruption of Mme. d'Esperance's trance were
fraught with such consequences to her health,
that it would not be advisable to encourage
anyone to repeat such an experiment.

If I am right in these suppositions, stages may
be constructed in the materialization phenomena,
starting with the most elementary visions and
ending with the perfected forms which, to the
uninitiated, can, perhaps, be hardly distinguished
from normal organisms. The higher the degree
of materialization, the harder the distinction,
the more perfect and stable the new formation.
Whether these can be of permanent character
remains to be proved. Crookes says that he was
allowed to take a lock of hair and scrap of clothing
from the materialized form. (If so, where are
they, and who possesses them now ?) Probably
in the higher stages of materialization some
dissolution of the material form of the medium
ensues.

A close comparison can be made between
materialization processes and creations by God.
The former almost seem to be a faint reflection
of the divine creative power, which is able to
evolve forms of far greater consistency and
duration. The creations of God are not tran-
sient, but remain until He Himself recalls them
into non-existence. The creations of the

materialization-mediums are quite fleeting and
do not last longer than the state of trance of the
medium, whether we assume that they are composed of the matter of which the organism of the
medium is itself composed or that they are " new
creations " of matter or material substance.
Whichever view is true, it may be that they
afford us a glimpse into the creative power of
God, for we cannot help imagining that the
creation of the world originated from the thought
of God in just the same way as materializations
are evolved by the thoughts of the medium.
What remains problematic is the part played by
the vital forces.   From whence do they come ?
Are they an independent group of world factors,
or are they also creations of God which are misused
by the mediums in the materialization process,
or is it more reasonable to assume that they too
only come into existence through the creative
action of God and of the mediums ?

ADDITIONAL REMARKS

The publication of a second edition has induced me once again to examine exhaustively
Schrenck-Notzing's first great publication, the
criticisms of Matilda von Kemmitz and von
Gulat-Wellenburg, and Schrenck-Notzing's reply.
I do not doubt that every reader, who (as is so
often the case) has only seen Kemmitz's pamphlet,
will be convinced that the whole thing is a fraud,

and that Schrenck-Notzing is a man utterly
devoid of all critical faculty. But if anyone
takes the trouble to work through Schrenck-
Notzing's book and his reply to his critics, he will
be compelled to shake his head over what can
only be described as the indescribable superficiality of the pamphlet of this lady doctor,
who had, indeed, but just taken her degree.

It is also not correct to say as Dessoir does,
that except Schrenck-Notzing, I can only refer
to Geley as vouching unreservedly for the
genuineness of the phenomena. This has been
done also in printed statements by Professors
Richet and Boirac, as well as by Dr. Bourbon
and de Fontenay, who was an expert on his own
account, and they did so as the result of sittings
in which they took part. Also Professor Courtier,
Clarapede, Bennet, Flammarion, etc., have
publicly declared that their conviction is the
same.

(" Psych : Studien." May, 1920).

Schrenck-Notzing has sent me, upon my
request, the letters which he received about the
investigations of the Committee of the Psychical
Research Society from Fielding, Fournier d'Albe,
Mme. Bisson[1] and Eva C. The essential parts
of these letters are printed lower down.
Fournier d'Albe declares that he is himself

---

[1] It has not been possible to obtain leave to publish these letters
in this English edition—Editor.

completely convinced of the genuineness of the phenomena, and that an amateur conjurer who took part in the sittings said that it was impossible to imitate them by trick methods. Fielding holds that a convincing proof of the genuineness of the phenomena was not achieved. The only possibility for fraud lay, in his opinion, in regurgitation, but this he regards as very unlikely. W. Whately Smith also says that the phenomena which he witnessed at six sittings in London could only have been produced by rumination; that it is not easy to see how rumination was the reason, and that he, upon the whole, has come in consideration of all the evidence for the case to the conclusion that it is genuine. ("The Psychic Research Quarterly" I, 3, 1921.) Since the sittings Schrenck-Notzing reports that a radiographic examination of the œsophagus and stomach of Eva C. has been made. The result is to show a normal condition of these organs unlike that of persons gifted with the power of rumination. Ruminants have always a distension of the stomach and, in order to achieve their results, have to swallow liquids in quart quantities.

Mme. Bisson, whose letters do not leave an unfavourable impression, complains that the methods of investigation in London were not at all adapted from the psychical point of view to the results which were desired. Instead of being

content with taking necessary precautions, conversation went on continually without any
restraint in the presence of the medium about
possibly undiscovered methods of cheating, and
the medium, as well as Mme. Bisson, naturally
became increasingly irritated. Moreover, right
from the beginning an unfavourable predisposition clearly prevailed in the minds of the committee, and this could not fail to have had an
unfavourable effect on the medium, and to
prevent the phenomena from reaching their
proper development.

It is a very common fact, which again and again
recurs, that persons who have taken part in an
investigation are convinced of the reality of the
phenomena during the sittings and immediately
after them, and record their opinions to this
effect in writing, and then after more or less of
an interval, as memory becomes fainter, they
begin to doubt again and explain the whole thing
just as decidedly in the opposite sense.

In the same way everyone on first making
acquaintance with parapsychological literature
undergoes a similar experience. As long as he is
under the impression of the recorded observed
facts he is more or less convinced, and then as
time goes by he becomes uncertain again, and
ends up by regarding the whole thing as a swindle.
If he again takes up the literature on the subject
he repeats the process, and this happens so often

that we at last become conscious of it and have to resolve to hold fast afterwards to the judgment which we have formed whilst actually engaged in considering the facts. This seems the only methodological point of view which is tenable in the circumstances.

From the theoretical point of view it would be extremely important, if the theory of observation could only be more closely worked out with the object of laying down rules for saying when an observed fact can be regarded as really supernormal and when the possibility of deception is excluded. The discussion about observed facts still continually breaks down over the impossibility of coming to an agreed decision of the question whether, given the conditions, the possibility of fraud was left open or not. It is necessary to lay down fixed criteria on this subject, or we shall never make any advance. The mere statement that " very probably there was fraud, after all," can always be made, but it cannot all the same claim to be always accepted as a valid argument, for if it were, the positive ascertainment of facts would be impossible.

In the same way explanations which are obviously impossible must not be allowed to stand. There is a kind of criticism, which is not criticism, which would rather admit the most senseless hypothesis than the existence of a parapsychic fact. One of the crassest examples of this may be found in Lehmann's well-known book on superstition and magic. He "explains" a report of Seiling about a partial dematerialization of Madame d'Esperance, by supposing that that medium "stuck her legs and, perhaps, the whole of the lower part of her body, through the opening of the back of the chair on which she was sitting," and he, himself, gives the measurements of this hole as nineteen cm. high by twenty-nine cm. broad. If you compare this measurement with that of a grown woman with her clothes on, you can only be astonished that a reputable investigator regards the proceeding as possible, and goes on to assume that the people sitting near-by noticed in no way how the medium got up and forced her legs and lower body with her clothes backwards through the opening of the chair.

The phenomena observed with Eva C. have recently received indirect confirmation through similar phenomena recorded in Crawford's posthumous work, "The Psychic Structures of the Goligher Circle" (Watkins, London, 1920), which reached me for the first time as I was correcting those additional remarks. (Schrenck-Notzing has given an account of the book in "Psych : Studien," August, 1921). These phenomena have upon occasion the same web-like formation as many of the structures produced by Eva C.

The genuineness of Crawford's work is vouched for, not only " by this one man," as Dessoir says. As regards the levitation phenomena, Professor Wm. Barrett has given a confirmatory statement. (In proceedings of S.P.R., Vol. 30, page 334). Crawford, in 1920, explained his unwillingness to allow the presence of witnesses as due to his fear that they might spoil the development of his medium by subjecting her to a bad psychical influence.

Moreover, Fournier d'Albe, who took over the further investigation of the medium soon after Crawford's death, has declared in favour of the genuineness of the phenomena. He writes to Schrenck-Notzing : " The phenomena are very strong and begin after 10–20 minutes. I have taken some photographs which show a quite regularly webbed-woven structure like chiffon. How it is produced future investigation must show."[1] F. M. Stevenson in the same way has confirmed Crawford's conclusions as the result of his own investigations and of the photographs which he took. (" The Psychic Research Quarterly," Oct., 1921.)

[1] Fournier d'Albe went back on this as the result of further investigation, and concluded emphatically that the phenomena were due to fraud. —Editor.

Extracts from letters about the London sittings with Eva C.

Fournier d'Albe to Schrenck-Notzing.
    " London, June 24, 1920.
" I was present at six sittings, of which four were negative. This seems about the usual proportion in London. The best sitting was last Thursday. Eva had the veil on, and not less than four different phenomena showed themselves one after the other in front of her face, but inside the veil—a finger,—a kind of veil,—a cravat with some lines upon it and a (paper) surface upon which the lines of a face were drawn. This last was grey and flat, and was fastened to her nose. All the phenomena were shown for some seconds. Mr. Dingwell the amateur conjurer of the committee assures me that the phenomena could not have been produced by trickery.

" Yesterday the result was similar, although Eva, shortly before the sitting, had had a cup of tea and a cake—a precaution directed against the regurgitation hypothesis. I was not present.

" The psychological effect upon the committee is the usual one. All the conditions, one after the other, were complied with, but, instead of being content they go on thinking out new ones. Whatever the reason, the phenomena are all of small measurements, and I think it probable that in London the larger phenomena will not be forthcoming."

"June 30, 1920.

"The London sittings are at an end, and the ladies journeyed back this morning to Paris. A total of about thirty (really forty) sittings took place and I think only about eight (really thirteen) were positive. I attended ten. Of these four were positive. The best was last Saturday from four to seven o'clock.

"We felt the cold wind (for the first time) and saw (1) a black string between the hands, (2) which changed itself into a grey membrane, (3) and then into a flat surface of grey felt (which was examined with an electric torch), (4) a mass which hung from the mouth like a stalactite; (5) later (2 minutes) a face drawn in natural colours upon a thin substratum (this was submitted to the light and photographed), (6) fibrous substance between the two hands which were held by D. This vanished suddenly before our eyes after I had examined the medium's mouth, (7) a 'finger' coming out of the mouth (photographed).

"This I am told was the best sitting. Since it took place two sittings were held on Monday and Tuesday without result. Mrs. Feilding was present at them. She seems to hinder the phenomena by her too critical attitude. Although

Mr. Feilding says that nothing took place which could not throughout have been produced by a conjurer, yet the personal judgment of all who took part is thoroughly favourable to the two ladies. We had hoped for phenomena on a larger scale, but the circumstances, especially from the psychological side, were not favourable enough for that."

Feilding to Schrenck-Notzing.

" London, July 17, 1920.

" Madame Bisson left over a fortnight ago, after a stay of more than two months, during which we had about thirty séances. The results, although very interesting, were, unhappily, not as important as previous reports had led us to hope for. That is to say that, inasmuch as the regurgitation theory is the only theory that can hold the field in opposition to that of supernormal ideoplasm, it is a pity that we never got phenomena big enough to warrant us in declaring, as a matter of scientific certainty, that this theory is insufficient, however great its improbability may be. I am extremely sorry that Madame Bisson seems very dissatisfied with our way of running things. All the same, I assure you most positively that her dissatisfaction is really baseless. We felt from the outset that it would be impossible that in a short series of experiments we should be able to add anything of value to the study of the phenomena from a scientific point of view. Madame Bisson has a far better installation in Paris than we have here for a study of this kind. We limited ourselves, therefore, to trying to confirm your and her opinion about the authenticity of the phenomena ; that is, to establish the absence of any trick or fraud of any kind, so as to prepare the English public for the forthcoming appearance of the translation of your book, the interest in which will obviously depend on the confidence which this public will have, that the facts described in it really deserve serious attention. There has already been a considerable polemic in the reviews and in certain books tending to show that it is all nothing but the merest humbug. At the least, therefore, a favourable report by the S.P.R., which is known for its caution in matters of this kind, would have a far greater practical value than an incomplete study of the nature of the ' substance ,' which is apparently what Madame Bisson wanted. Madame is wrong in thinking that we began the experiments with the conviction that it was all merely a trick to be shown up. But as we knew that the chief thing would be to be able to give an effective answer to the question ' trick, or genuine phenomenon ? ' we included in our committee certain members also, who, from their knowledge of conjuring, would be able to speak as experts from that point of

view. They were not, however, in any way con-
jurors by profession."

Feilding to Schrenck-Notzing.

"Rockport, Ireland, August 28, 1920.

"I think I must have expressed myself badly
when speaking about the regurgitation theory.
I never intended to say that we had adopted this
as the most probable theory. I only said that
the phenomena that we saw in London were
not large enough for us to be able to say that
they, taken alone and without considering what
had been observed by our predecessors, rendered
the regurgitation theory impossible."

Eva C. to Schrenck-Notzing.

"London, May 5, 1920.

"The sittings here seem likely to go well.
Out of five, three have produced something. I
very much hope that the results will be altogether satisfactory. After this I believe that
I shall have done my duty as regards these
sciences. At all events I shall have done all that
was humanly possible for me. I have often
needed courage, for all these questions are very
distasteful and not always very considerate for
the medium. The control is always very painful
to me, for I am really a bit of a savage, and all these
exhibitions of my person are really hateful. I
accept every condition, and desire to do so, as
I understand that the whole interest in these
phenomena is based on control before and after.

Madame Bisson has recently published the following statement in " Psychica," 1921, May 15, No. 3 :

" The first five sittings in London gave excellent results. But the experimenters were not used to materialization phenomena and the room for the sittings adjoined a very noisy room, from which strong disturbing noises made by messengers kept coming so that the medium was woken up and made nervous. The hour of the sittings was changed to later in the day, and this was very inconvenient for the medium who only got home late after tiring sittings.

" Yet in spite of all distracting causes we got some good and interesting manifestations. One day, both hands of the medium being held by one of those present (a conjurer) a parcel of substance appeared between this gentleman's hands and Eva's. This substance grew bigger on their hands and then developed. A small woman's face appeared. Mr. D., the conjurer, cried out : ' She has blue eyes and red lips— she is smiling to me.'

" I do not yet know the report of the S.P.R., and I shall avoid commenting on it. I got a letter to say that Mr. D. had delivered a lecture on our work in which he is said to have declared that he could not admit that there was any fraud, for fraud would seem to him more extraordinary than the phenomena itself."

CHAPTER VI

THEOSOPHY—RUDOLF STEINER

THE superabnormal psychical and physical phenomena which we have so far reviewed form to-day the field of research for the initial stage of parapsychology, the newest branch of philosophy. The beginnings of these investigations—not yet recognized by many of us as an independent scientific branch—correspond roughly to the beginnings of hypnotic research, a further point of resemblance consisting in the fact that though inaugurated some decades ago, they were forgotten later, only to be revived once again towards the end of the last century.

Just as the phenomena of hypnotism are primeval and are to be found right through history down to the present day, so, too, are the supernormal phenomena with which we are concerned in this book, even if we are still unable to distinguish between truth and fiction with regard to the traditions about them in history. To make this distinction will only be possible when we are on a firmer footing all round than is the case at present. It is certainly probable that, ten years hence, we may look upon much in history—

particularly in the history of religion—with
very different eyes, though even so, there will be
quite enough errors, fraud and superstition left.
Parapsychic phenomena have at all times served
as the starting point for an immense amount of
premature metaphysics about the soul.   The
first and easiest interpretation of many of the
phenomena is the spiritistic. It consists in the
assumption that it is not the psyché of the
medium which develops supernormal faculties,
but that departed spirits establish themselves in
the organism of the medium, manifesting their
presence through it, and even substracting
organic substance from the medium, in order to
be reincarnated or materialized. This spiritistic
conception which the latest observations of
materializations have rendered so extremely improbable makes so great an appeal to the naïve
observer, that it has existed ever since parapsychic phenomena became known, though the
same significance was not always attached to it.
The last great upheaval started towards the
middle of the last century, when so-called
" table-turning " came over from America. This
was the birth of modern Spiritism, which soon
spread over the whole world, and even to-day has
its circle of adherents everywhere.   But as every
new wave of thought—whether high or low—
sooner or later loses its native force, so, too,
spiritism saw its strength wane.

As a matter of fact, spiritism to-day is no longer the most modern form of occultism. The strong religious tendency of the present day working on occultism has produced a new movement, which in contradistinction to Occultism shows a decided religious bias, namely that of Theosophy. In theosophy spiritism has to a certain degree been merged, for spiritism has no tenets which Theosophy does not recognize. Theosophy, however, goes beyond spiritism, accepting second sight, telepathy, visions of spirits and materializations as of ordinary, everyday occurrences—manifestations of a lower grade, which, indeed, it treats as real facts, but no longer as of primary importance.

Theosophy desires to offer more. Its aim is to be able to unveil those mysteries of the universe into which spiritism dared not probe. Theosophy believes it will be able to make the higher knowledge, by which time and space cease to exist, accessible to all. Theosophy considers itself as a sort of godhead, above all philosophic thought, which is but a preliminary step, and which it contemplates from a higher angle—like Hegel when he, from the philosophic point of view, looked down on mere rational discursive thought. Theosophy believes itself to possess a more advanced knowledge, though despite this, it professes no close relationship with traditional religious beliefs which depend

upon revelation. It is much more like a religion
in its initial stages than like a creed, which has
existed for ages and already follows a steady
course. For long confined to a comparatively
small, exclusive circle of adherents, it has grown
considerably of late years, particularly since the
German revolution, as is indicated by the considerable increase in theosophic publications.

Theosophy is not a spiritual movement of a
kind peculiar to the present day. In earlier
epochs, similar movements existed which sought
for loftier religious understanding, beyond the
limits of philosophy and traditional religion.
Throughout the Middle Ages this tendency lay
hidden beneath the religious life. It was known
as Mysticism. It held that an unmediated
elevation of the soul to God is possible, which
would lead to the unification of the human
personality with the Divine Being, to a *unio
mystica* or *deificatio*, or as German mysticism
terms it, a " Vergottung." As compared with
this older form of Theosophy, which still lives
on in Catholicism to this day, modern Theosophy
emphasizes the intellectual side. It is based on
knowledge although inner religious depths are
not lacking in it either.

The whole movement originated in India and
England, and even to-day the English Theosophical tendency is the dominant one. At its
head stood the Russian Blavatsky, possibly herself

a medium, though at the same time an extremely cunning impostor and as fascinating a personality as only a hysterical Russian of that class can be. The report upon her by the psychologist Hodgson, who was deputed to make investigations on behalf of the Society for Psychical Research, was as unfavourable to her as it could be.

At the present day also, as at its origin, English Theosophy is dominated by the mind of a woman, that of Mrs. Annie Besant. Morally, she must be considered much higher than her predecessor, who at one and the same time seems to have fulfilled the functions of Russian political agent, and mediatress of the higher revelations. Mrs. Besant is no impostor, and makes no pretensions to an exalted position among her fellows as did Mme. Blavatsky. She must be an imposing personality, and not only as a public speaker. According to her own confession to James, years ago, she is characterized by extreme diffidence in personal intercourse, though recklessly brave in public, but Keyserlinck, who visited her in India, was quite differently impressed. " Of one thing I am convinced," he writes : " this woman controls her person from a centre in a way that I have seldom seen equalled. . . . Mrs. Besant has her abilities, sensations, will-power, so in hand, that she appears capable of greater efforts than those who are more highly gifted." The literary productions of Annie Besant have

nothing distinguished about them.   To a scientifically inclined reader her books are quite unpalat-
able.   As a critic once put it :   " Should an
exhibition ever be held of the malformations of
human thought, there would be a rush to be the
first to get her books for it."

The centre of English Theosophy must be
looked for in India, where the speculations of the
old Indian Theosophy are deliberately renewed.
The reader is overwhelmed in Annie Besant's
books by a mass of Indian expressions, and
doubtless the effect on half-educated people is
often to create a curious haziness of mind, an
atmosphere of mystery, full of wondrous thoughts,
favourable to any kind of auto-suggestion.   Not
only do the ideas of English Theosophy emanate
from India, but the close relationship with that
country is proclaimed by the fact that the
headquarters of the movement is in India itself,
in Adyar, where Mme. Blavatsky founded her
Indian Theosophist Society in 1875.   Annie
Besant is now its president, and has made India
her permanent home for many years past.   The
fusion of race and the ideal of indiscriminate
brotherly love have in this movement not been
confined to mere expression, and the British
members of the Society are even in sympathy
with the Indian national movement.   Mrs. Besant
took a leading part in the last Indian National
Congress, and the Anglo-Indian authorities have

been faced with difficulties as a direct consequence of her action.

The Theosophical movement has extended from England to Germany, and Annie Besant herself came over to give a series of lectures. Some of her voluminous writings as well as those of her disciple Leadbeater have been translated. The prophet-in-chief of German Theosophy to-day, however, is Rudolf Steiner of Stuttgart. He is the son of a small railway official, and was born in Hungary in 1861. Even now, he still feels himself to be the " son of the proletariate," and the " proletarian " seems to him the greatest historical reality of the present, whose mission it is to lead humanity on to a higher level. I once heard him lecture on state reform : " The Three Limbs of the Social Organism." He is a tall, dark person, with an intelligent expression. He enlarged for two or three hours, in a voice of remarkable carrying power, on very few ideas ; or, rather, he actually bellowed as though he meant to drown any possible contradiction. His lecture was extremely monotonous, and at the same time very vague, a typical German revolutionary programme of reform. The enigma as to the why and wherefore of his hold on his following, and how it is that hundreds, nay, thousands, believe in him, is still unsolved. Neither is it easy to understand how it is that a group of women are said to have pursued him from place to place at that time, in order to hear him speak again and again.

The fundamental idea of Theosophy is expressed in the belief that the world of our senses does not represent the whole of reality ; but that higher spheres exist, and that mankind is enabled to gain insight into this higher world by reason of second sight òr " clairvoyance." That which to us is the world forms but a small section of the actual universe of being.

The world is not alone in possessing unknown spheres and gradations ; what is true of the world is true also of mankind. This new interpretation of man's constitution is considered by Steiner to be of such importance that he calls his whole train of thought Anthroposophy instead of Theosophy. This change of name synchronizes with a slight change of emphasis from the religious to the intellectual sphere. Steiner repudiates the dominant conception—that man is composed of two parts : the body and the soul. He admits of no less than four component parts. Over and above the physical body, which is made of the same consistent parts as the inanimate world, is (2) the " living or ætheric body," which is practically represented in neovitalitism, by entelechy, psychoid, vital factor or the like. The word " æther " (body) has no connexion with the " æther " of physics. The (3) third " member of the human being " is the " sensive or astral body," and transmits suffering and desire, joy, lust, passion, etc. The (4) fourth factor, mankind's own peculiar prerogative, is the "ego," the "body of the ego," where, again, the word "body" is not to be taken literally. Plato would have called it "reason" instead. The sensitive body is said to be formed in the shape of a "longish egg," in which both the physical and ætheric bodies are embedded! It ranges over them in every direction like a "form of light." We meet here correct opinions strangely transformed and intermingled with phantastic ones. The semi-materialism is characteristic, for though Steiner is at pains to speak of "body," he is still quite aware that he is employing a false term.

The belief held by all Theosophists in reincarnation is of the greatest significance. The idea of the transmigration of the soul is alien to the European spiritual world with but few exceptions—Pythagoras, for instance. It has never gained ground in Europe to any great extent. This idea, too, originated in India, like all other fundamental Theosophist teachings, and there, for thousands of years, it has formed the main part of all religions. It is a thought which, if taken in a deeper sense, can markedly increase the sense of responsibility in man by the belief that his future fate is dependent on his present life. It also confers a strange peace of mind on some individuals, when combined with other beliefs. When life is regarded as lasting for a very long time, time ceases to be precious ; there is no necessity for a hurry ; in future existences there will be opportunity for everything now denied.

The whole development of culture is visualized by Steiner as dependent on the ever-increasing influence of the ego on the remaining parts of man.   Indeed the secret teaching of Theosophy consists in an ever greater control of the ego over those parts.   Man becomes master of his character, his passions, and, at the highest stage, even of his physical body.   He controls the circulation of the blood, as well as his pulse.   In this it is easy to recognize the influence of the ancient Indian Theosophy in which, on reaching the highest stage of development, man attains mastery over his organic functions.

But this is not yet the heart of the new secret teachings.   The final aim is the achievement of a higher knowledge—that of so-called " second sight," the contemplative observation of all the profundities of reality.   All mankind can attain this understanding, when, through unceasing mediation, they release those higher faculties which lie dormant within them.   To the " seer " nothing would remain hidden.

It is amazing in the extreme to hear of the results attained by means of the higher mental faculties, which, though dormant in most, appear thoroughly awakened in Steiner. We learn of the most unheard-of things concerning the happenings of the universe. What are all the attainments of geology and astronomy in comparison to the visions of the Theosophists ? The past of the solar system and of the earth are relentlessly unveiled to us. We hear of new ages of which no one knew anything. We learn how mankind was formed in the " lemuric " period on that continent which existed between Australia and India. Further, Steiner alludes to powerful spirits who wandered about earth before mankind existed, and he knows of whole cycles of culture which existed in prehistoric times. All Plato's descriptions of Atlanta read like a dry and harmless report when compared to Steiner's clairvoyant visions. Even Helene Smith's Martian and ultra-Martian apparitions pale before them. Every now and then are interspersed glimpses of angelic figures, higher spirits, which were active in the earlier stages of the solar system. Even a pre-historic Christ is not missing from the cosmic birth of the world, when earth and sun parted. This Christ, or Sun-Man, is said to have instructed seven great teachers—the teachers of the ancient Indies, though these ancient Indies must not be confounded with the present geographical conception of India. There is said to have been a " higher "

pre-historic India.   But in that case, one asks in
vain why it is still called India ?

Even Schelling's later reckless Theosophical and
cosmogonic speculations are child's play in
comparison to Steiner's "inspirations," which
might better be thought to bear an analogy
to the philosophy of later classical times.   In my
opinion, however, even this comparison is too
pale.   I can find nothing better to compare than
the Apocalyptic Scriptures—the Revelations of
St. John—or, better still, the apocalyptic visions
of David of Lazareth.   True it is that Steiner
lacks the touch of power which is found in them,
but like them he sweeps grandiosely above all the
probabilities, and like them deals with the
milliards of years, æons and super-epochs.   The
question arises—if the analogy be pursued—is
Steiner then also mentally deranged ?   The
contents of certain of his writings certainly seem
to suggest this.   Or how otherwise can these
emanations of the spirit be understood ?   It
must be admitted as an "extenuating circumstance" that Steiner is not unique in this regard.
One need only examine the recent publications
in German of the works of Annie Besant, and of
her disciple, Leadbeater.   Here, too, allusions
are found to whole series of realities of which the
rest of us are entirely unaware; and in another book
an attempt is made to unravel the secrets of the
construction of the atoms by means of clairvoyance.

Who will take the trouble to compare all these conflicting revelations ? I fear they would prove as full of contradictions as the scriptural Apocrypha, though this fact would cause no discouragement to the Theosophists. They would merely say that some of the revelations were genuine—others only pseudo-revelations. Even if they were not full of contradictions, that would not prove them objective ; it would still be simpler to assume that they are relatively interdependent. The point so much against them is that they fail to show any interconnexion with the ordinary sciences. And yet any claim to objectivity which they might have must at least entail some sort of connecting link with established knowledge. A " clairvoyant " description of prehistoric facts, if genuine, should by rights have some connexion with prehistoric science, and reveal to persons interested in early history steps and stages of development which would bring to light the interconnexion of historically recorded facts. In short, it would do what genius does. We find nothing of all this in Steiner's conceptions. Further, the reader who is versed in psychology feels the absence of detailed descriptions of the nature of this process of revelation. We are overwhelmed by a mass of assertion—mere assertion—nothing else. And when the question is raised : Will the student of Theosophy himself reach the stage of personal inspiration, the answer is bewildering, for the reader is told that he is on the threshold already of spiritual enlightenment as soon as he begins to hear and understand Steiner's revelations. From the time when these revelations are received and believed, we are told that we have part in them and have ourselves received them. This is surely a dubious assertion, and might even be regarded as hallucination; for Steiner must surely know the great difference between mere belief in something and an act of higher intuition.

Steiner writes quite differently when he is on neutral ground. For instance, his essays on the earlier philosophers are thorough and impressive. We should expect him, therefore, to show a keener appreciation of the value of some psychological analysis of the insight which he claims to possess.

The path to second sight is reached by way of strange exercises of mental concentration—*exercitia spiritualia*—which are directly derived from the Indian school of spiritual training. In the first place, Steiner advocates a contemplative study of certain flower-like drawings. These are alleged to have a highly symbolical significance. Thus a black cross is the symbol of the baser passions and lusts; therefore, during meditation seven red roses are placed in the centre of a black cross to denote passions and lusts under control. After continued contemplative exercises, the student is stated gradually to get outside his own ego and to become conscious of the higher spiritual world. Unfortunately, records in literature of the effect of such contemplation and of its gradual increasing power are extremely rare. It is obviously a question of generating auto-hypnotic conditions, which create a favourable basis for auto-suggestions of every description. The suspicion arises, that the student of Theòsophy experiences nothing beyond an auto-suggestive strengthening of his faith in his Master and possibly a few corresponding hallucinations. But nothing much is achieved by such a conclusion. According to numerous reports from India, there can be no doubt that a protraction of such contemplative exercises has a peculiarly strengthening effect on the human mind. We are continually reminded of the fact that the Indian Yogi are renowned for their complete control over their psyché. Keyserlinck is also of opinion that Mrs. Annie Besant's self-mastery is attributable to her Yogi exercises. And surely the deep satisfaction which has been attained by many adherents of the Theosophical movement, and which I am sure exists from the statements which they themselves have made to me, is attributable to these contemplations. A closer study of the whole subject has become a pressing duty, and may have important results for the self-education of mature minds.

A deeper study of these conditions of spiritual concentration is, however, hardly possible in Europe. I am almost inclined to believe that all that has been achieved by exercises in concentration as practised in Europe has reached only to the fringe of what has been done in India. The whole mind of the European is far too active and too engrossed with worldly affairs. The European is not able to devote himself to spiritual exercises in the way the Indian can.

The study of Indian self-absorption must be carried out in India itself. That is why it is of such great parapsychical importance, for, if we are to believe the accounts of travellers to that land, it represents a method which systematically obtains complete mastery over those parts of the organism which are not subjected to the conscious will, as well as also furthering mediumistic faculties. European mediumship is the gift of chance—certain persons evince abnormal parapsychic phenomena, we know not how or when. In India the problem of the methodical production of such faculties has apparently been solved for centuries. This assertion might have been ignored so long as there was any question of the reality of the parapsychic phenomena. To-day, when doubt is no longer possible, the Indian reports have also become of interest to us. Whatever the outcome, they deserve to be investigated, even at the risk that the mystery may remain unsolved.

Indian Theosophy has this in common with European Theosophy : the majority of its adherents are believers only—very few have real knowledge. " The majority of those that I have talked to," said Keyserlinck, " believe only, though a few are convinced that they also know, and report to me as naturally and calmly about unheard-of events, as a naturalist on his latest discovery."

All that we have so far heard about Steiner and his teachings does not, however, solve the problem of his ever-increasing influence even on many people who are of a superior moral character. It is evident that these weird occult teachings, which promise an insight yet undreamed of, must have a mysterious attraction for many. But the deeper satisfaction found by so many in Theosophy is surely attributable to the high moral tendency which permeates Steiner's teachings (though it easily is lost sight of by persons who stand outside Theosophy in the face of his strange metaphysical teachings) ; which pervades his whole theories ; and which brings Theosophy into closer relationship with Christianity than seems apparent at first sight. The real reason for Steiner's great influence, therefore, lies in the higher values interspersed in his metaphysics which ensure him respect, confused though his speculations seem. To this must be added the fact that Steiner rejects certain forms of older Christianity, e.g. its contempt for health and strength—which are alien to the modern mind. In this, again, he concurs with Indian views, by which Yogism and the highest condition of concentration are confined to those in perfect mental and physical health. Steiner's writings insist on inner mental health, and he requires this condition from the students' of Theosophy. In far-seeing guise, Theosophists believe that a religious conception of life is essential to the healthy life of the soul. Work and devotion are for them the pivots of life.

But all this does not prevent the rest of Steiner's writings from being extraordinarily confused and muddled quite apart from the fact that little reliance can be placed on his terminology and expressions. He does not consider it necessary to confine himself to the use of words in their ordinary sense and what is still worse, his own terms are themselves hopelessly jumbled together.

ADDITIONAL REMARKS

Since the first edition of this book was published, the Theosophical movement has spread still further. Its claims are greater than ever, and the number of its opponents is also increasing.

I have avoided taking any part in the platform campaign about it. The effort to find the truth, in which I should like to help, will not be for-warded—it will rather be hindered—by the controversies of public discussion. As I am told, Steiner in his new form of polemics has attacked me personally, but I do not know how or when and I shall take care not to be tempted to answer insult with insult.

The two main points, which most urgently ought to be cleared up, are first as to the nature of Steiner's so-called "second sight" (Hell-sehen), and secondly as to the question (which is not without connexion with the first) how meditation is practised in the inner anthropo-sophical circles, and how in Steiner's own case it first came to fulfilment, for Steiner (as is known) comes from the school of Mrs. Besant—a fact which anthroposophical circles generally rather markedly avoid mentioning, as they prefer to regard Steiner as a man who has never been spiritually indebted to any other.

It is further worth noting that the expression "second sight" is used by Steiner in an unusually wide sense. Generally the word is used to denote the capacity of perceiving a thing although the normal condition of sight are not present, that is although no light waves from the thing can reach the eye because the thing is too far away or hidden somewhere. Steiner, however, means

by "second sight" the pretended higher
functions of the soul which are dependent on
the so-called astral body, through which the
Theosophist is able to read the "Akasha Record,"
that is to see the impressions which are retained
by a higher actuality of all the events of our
homely earth.   Yet all this must, he says, be
treated merely as a use of metaphors, both when
we speak of "impressions" or of "reading" them.
If this is so, it would be better not to use the
word "second sight" at all.   But we have to
reconcile ourselves as best we can to Steiner's
habit, not of coining new words which could be
used in a definite and recognizable sense, but of
taking old words and using them in a new and
altered meaning.

The deciding question, with which his whole
standpoint stands or falls, is the question whether
"second sight," in Steiner's sense, is a fact
which actually occurs or not, and whether he
himself possesses it.   If this is answered, then the
question of the existence of the Akasha Record—
or at least, the question whether it is capable of
being proved to exist will also be settled.

The obvious idea of making experiments in
the ordinary sense of the word with the faculty
of second sight leads to no result, since the
reading of the Akasha Record is not a second sight
of the kind which permits of experiments.   The
higher degree of second sight has got to be

tested by itself. This test can naturally not be carried out by any proof of the truth of Steiner's deliverances on the subject of the early history of the sun's system or of mankind. We have no methods of verification which are not dependent upon the usual scientific considerations about the sun or mankind. We cannot help from the scientific point of view regarding most of his stories as bottomless imaginings, and the most that we can possibly do is perhaps to discuss one or two of them, for instance, the question of the Atlantes, or of the Lemurians. To this Steiner retorts that our own scientific interpretations are merely uncertain hypotheses. A decision can therefore only be reached if Steiner is himself so kind as to help us ! He must give us information from the Akasha Record about past events, which we do not yet know, but which we should be able to verify with reasonable certainty by the use of the normal channels of knowledge.

Accordingly, I hereby challenge Steiner personally to give us an opportunity of verifying in this way his assertions about his superhuman capacities. And since according to views current in Theosophical circles, which Steiner also shares, as the mind rises to higher levels which are still levels common to clairvoyant faculties in the ordinary sense, it develops a power of getting free from the body as well as other capacities, it is desirable that the opportunity for verification should extend to the whole of the phenomena under consideration and especially to the capacity which Rittelmeyer maintains is possessed by Steiner of seeing the contents of other people's minds. Steiner is said to have already once put himself at the disposal of Kulpe in Munich for a psychological examination, but Kulpe unfortunately declined the task for want of time. I may, therefore, very naturally express the wish that he will again make the offer—especially as he declares that he has himself a theoretical interest in psychology.

Some of Steiner's supporters have been very much offended because I said that according to Steiner, his followers share in his revelations as soon as they take his assertions to themselves and believe in them. It is (according to these supporters) characteristic of Steiner that he does not demand belief but only that we should follow his thoughts. On the other hand it should be noted that he expressly declares that the study of his revelations is itself a means of " reaching knowledge on one's own account," indeed, that it is indispensable thereto. " In all esoteric training such study acts as a preparation. If a man tries every other means and does not take to himself the teaching of his esoteric guide, he will not reach his goal." For this teaching is not mere words, but a " living force." This can only be described as meaning that the esoteric student has got blindly to trust himself to the truth of Steiner's revelations so that they may become his own living convictions.

Moreover, the requirement of humility and of the denial of the self naturally works in the same direction to produce a credulous acceptance of the doctrine.

Again, how did Steiner, or whoever was the first and original " clairvoyant," attain to his capacity of higher vision, if the study of his revelations is indispensable ? There must have been a time when revelations did not exist which could be studied. And we should like to know what is the actual result in the circles of Steiner's followers of the practice of concentration with a view to achieving "second sight." I have not yet been able to find out that any clairvoyant worth mention has arisen among all his numerous followers, some of whom are adepts of distinguished ability.

I should further regard it as extraordinarily important to establish once for all, by a special investigation, how far the contents of Steiner's clairvoyant revelations correspond in detail with those of the English Theosophists, especially Mme. Blavatsky and Annie Besant. I have not so far being able to find leisure to make the comparison and can only urge others to do so. Such a comparison is necessary for the purpose of explanation. The correspondences between the revelations go so far that after a renewed consideration of all the factors I no longer think it probable that we have to do with some mental derangement in Steiner's case. It seems much more probable that the principal cause is not madness, but merely ideas derived from Mme. Blavatsky. The delusion of madness and of suggestion cannot be at once distinguished. Both have this in common, that the person holding them does not hold them for logical reasons and withstands and is unaffected by all arguments to the contrary. It is only by tracing them to their origin that we can decide definitely whether such delusions are due to madness or suggestion. I will hazard the conjecture provisionally that with anthroposophists the fundamental conceptions become implanted in the mind during states of meditation which are very like hypnotic states. That the motive power is suggestion seems very clear from the statements of Frohn-meyer of the school of Annie Besant, for he implies that both definite hypnotic suggestion as well as telepathy are employed. It is true that he means by this the employment of tele-pathic suggestion, and it is unfortunately true that of this there are so far very few properly recorded observed instances, although it seems true that it is a fact and, from the psychological standpoint, badly needs closer investigation, since it probably makes possible a not inconsiderable part (which has so far been ignored), of deceptive pseudo-mediumistic performances. The fact that in Mrs. Besant's following unquestioning mental obedience of the kind demanded by the Jesuits is required, makes it doubly probable that Steiner found his fundamental conceptions in this way. This, of course, does not exclude the possibility that he possesses parapsychological gifts, either by natural endowment or as the result of self-education by way of meditation. But if we are to believe that this is true, it is for him to furnish us with proof.

THE SCOPE FOR NEW INVESTIGATIONS

WE have now concluded our investigation of some of the domains of modern Occultism. Many may be disillusioned who hoped to see the portals of the higher world, —that of life after death,—flung wide before them at last. This hope has not been fulfilled. The strange, impenetrable wall, which hinders us from casting a glimpse into the reality beyond the grave, will not permit itself to be opened,— not even by means of the phenomena of Cross-correspondence. It is as though we are deliberately meant to be kept in the dark with regard to what awaits us. There are no proofs whatever which force us to the belief that there is any spirit responsible for the productions of mediumship other than the spirit of the medium—him or herself. No matter how high or low the productions of mediumship may be, they must still be ascribed to the medium, for the unconscious or somnambulistic productions of the mind may very well be above or below that of the consciousness. In the same way, however great may be the resemblance between the character of the mediumistic reproduction and the character of the individual—alive or dead—portrayed, mere likeness cannot be taken to be a proof of identity. We know of no limits to the faculty of impersonation ¯which some people possess. Lastly, a medium's knowledge of facts can never be taken as a conclusive proof of the presence of a dead person, for the statements made have in the first place to be verified, and if verification is possible, it is itself a proof that these facts can be ascertained by other means than the direct memory of the dead person. The same, too, applies to materializations. Spiritism, therefore, cannot be *proved* by incontrovertible reasons. On the other hand, neither can it be *disproved* by incontrovertible reasons. A conclusion can only be arrived at, based on general impressions, which will be the more correct the less bias there is either of sympathy with or of antipathy to the fundamental idea of spiritism. And yet it is true that we cannot do otherwise than to see and judge of the spiritistic hypothesis by its relationship to our general conception of the universe. In the absence of conclusive proof, we can neither accept nor reject it without inquiry into its consequences.

Although, then, we have to conclude that the main metaphysical expectation, which is closely connected with occultism, is proved to be fundamentally impracticable, the further scientific results are all the more important. Amid the waste matter of vulgar spiritism, most remarkable psychic and psychophysic phenomena of supernormal character have been discovered. We are now on new scientific ground. Many things are still veiled in clouds, vague, and only recognizable in outline ; others are still completely hidden ; others, again, have been established with comparative certainty. It is no longer an open question whether we have firm ground under our feet with regard to these problems, or whether all is illusion, deception, and fraud. The assertions of eminent investigators—some among them scientists of world-wide renown—are too numerous and too decided. All[1] who have gone in for a systematic study of the phenomena have arrived at a positive conclusion to a greater or less degree. To ignore their combined testimony would be but unscientific, dogmatic prejudice. No other scientific attitude is possible than that of taking in hand the examination and verification of the results already obtained.

A criticism that goes so far as to refuse to make a closer investigation of facts which have been asserted, becomes pseudo-criticism, and no longer impartial, when the facts have been asserted by reliable observers. The attitude of a considerable number, particularly of the older pro-

[1] Including A. Lehmann, Henning, and Dessoir.

fessors of philosophy and psychology, is strongly reminiscent of those Florentine savants who denied the astronomical discoveries of Galileo, and refused to look through the telescope for fear of being convinced. It is crassly untrue to assert that all clairvoyants and mediums refuse to submit themselves to scientific examination. The scientific superficiality and general lack of principle, which characterizes some authors when it becomes a question of the real facts, is alone responsible for such an assertion. Equally superficial, and probably only due to lack of knowledge of the literature upon the subject, is the statement by Hopp to the effect that science has no means of investigating the problem of telepathy, but must confine itself to the examination of such cases as chance may offer.

It is a test of the intellectual worth of a scientist if he is ready and willing to investigate problems which, if true, may open up wide issues, or if he retreats with diffidence from all that might bring with it in its train a revolutionary change in his existing theories.

German research has in the first place to ascertain what has already been accomplished. It is now no longer admissible to regard the entire parapsychic problem as *terra nova*, on which no man's foot has stepped yet.

When Wundt declared that if parapsychology were justified there must be two worlds—the

first, that which exists in accordance with the
laws of Galileo and of classic mechanics; the
second, that of the gnomes, rapping spirits and
magnetic mediums, in which the laws which
prevail in the first are not in use—it must unreservedly be admitted (apart from the language
in which his idea is clothed) that this "second
world" lacks that transparent and reliable
structure which is possessed by organically dead
nature, taken by itself. This is not only applicable to parapsychic phenomena, but is equally
true both of normal psychology and of organic
physiology. Biology, also, is incapable of in-
dicating in advance the progress of organic
development. This unreliability is peculiar to
all non-inorganic parts of reality, and it is very
doubtful whether it should only be attributed,
as it usually is, to the complication of the phenomena. Such an explanation is hardly true of
mental phenomena, and does not very well
apply at all to the vital factors. The reason
really consists in the fact that there is no question
here of a number of separate entities, and their
connexion to each other, as in the case of the
atoms which go to build up the elementary
entities of the inorganic world. The mechanical
conception of the universe, on which Wundt
and all the other Parallelists base their philosophy,
has been found to be fundamentally false. It is
necessary to make a clean break with the mechani-

cal conception, not only for the purposes of theory, but also for that of practical investigation, and no longer to shut one's eyes to the metapsychical complexity of the problems in which we are engaged.

There should soon be no lack of suitable mediums with whom to experiment. There are many indications that individuals with parapsychical constitutions are not so rare as is generally imagined, though it may be true that they are a little 'scarcer in Germany than among the English-speaking and Latin races. On the other hand, a whole number of mediumistically inclined individuals have been discovered recently. It is to be hoped that at least a proportion among them—those of the greatest value from the point of view of the renewal of scientific investigation—will place themselves at the disposal of a science which no longer need assume that it only has to do with fraud, and which is ready to investigate on critically objective lines. It is for this reason that I insist on the need for further investigation. One of these mediums, on whom v. Wasielewski and Tischner have made their chief experiments, does not even appear to base her work on the spiritistic theory, so that the tests in her case could be carried out in entire freedom from the spiritistic atmosphere.

It is greatly to be desired that we may soon have the luck to get physical mediums to place themselves at our disposal in the earlier stages of their development, so that they may be withdrawn from the influences of spiritism. It might then, perhaps, be possible to obtain parapsychic and paraphysical phenomena in a form which does not clothe itself in the spiritistic garb, and is thus divested of the distasteful atmosphere which surrounds such phenomena at present. The possibility exists, however, that the further development of these phenomena may be dependent on the necessity for the very favourable auto-suggestive influences which emanate from the spiritistic beliefs of the mediums. If this be, indeed, the case, we must needs resign ourselves to the acceptance of parapsychic manifestations in this strange guise.

A serious difficulty in investigation is the aversion manifested in spiritistic circles to scientific research. This circumstance is not only due to the indifferent and uninterested attitude hitherto adopted by science in Germany with reference to parapsychological problems, but is also caused by an instinctive fear that a closer investigation might prove the claims of the spiritistic interpretation to be unfounded. *Mundus vult decipi.* In the summer of 1919, one of my audience in Tübingen informed me of a writing medium who seemingly appeared to be capable of quite interesting phenomena, and he

promised to put me into touch with this person—
a servant girl. His uncle, however, with whom
the girl was in service, declined to allow a scientific investigation. I was not even permitted
to obtain a glimpse of the voluminous automatic
writings. Nothing could show more strongly
how necessary it is that the medium's scientific
interest should be trained, or that the mediumistic
faculties of persons who have some scientific
interest should be developed.

The proper attitude towards the spiritistic
hypothesis can only be that of critical examination. It is clear that this theory can only be
regarded as proved when all the probabilities
attendant on it are also proved. On the other
hand, a refusal to associate with persons who
accept the theory might lead to the exclusion
of such investigators as Myers and James. The
fact by itself, that anyone accepts the spiritistic
theory must not—however unsympathetic it may
be to us—lead us to condemn that person generally
as unreliable so far as he is not shown to be guilty
of unreliability by his manner of dealing with
mere facts. This point of view should lead us
to a *modus vivendi* with supporters of spiritism
in Germany, just as a *modus vivendi* already
exists in England.

A more minute analysis of parapsychic phenomena is the great need. To accomplish this the
analytical methods of normal psychology must

be invoked in full measure. The number of fundamental psychological conceptions which the parapsychologists have hitherto used is far too small. I miss, above all, the distinction between acts of representation and acts of thought ; to say nothing of the omission of the analysis of parapsychic acts of thought. How, for instance, does a medium distinguish psychometric or telepathic thoughts from his " own " normal ideas ? Or does he not make any distinction ? Neither do the reports show clearly—as a general rule—whether the visions of the mediums, with which we are dealing, are, in their nature, hallucinations or only representations. In short, analysis is still in its infancy. What is really required is Parapsychologic Experiments combined with introspection. In all cases where parapsychic phenomena occur without the more profound trance conditions, this cannot offer very great difficulties. Tests must also be made with mediums in a state resembling hypnotic trance. Attempts must be made, where possible, to persuade them to practise introspection, and they must be trained in it. Above all, Vogt's method of artificially narrowing the consciousness must be applied, in order to obtain the greatest number of statements based on retrospection. In short, all possible means must be employed.

In parapsychology, all efforts must increasingly be centred on the attainment of greater objectivity in methods. The ideal would be to register mechanically the phenomena throughout their whole development, so that subsequent study in complete leisure might be possible. Too often the outward conditions in which the sittings take place make observation very difficult.

In this way only will the sceptic be convinced and compelled to accept the objective proofs afforded by photography and the registering apparatus. The extensive use which Schrenck-Notzing made of photography, the several exposures taken from different angles at once, and the use of stereoscopy mark a considerable advance in methods. But it would be desirable to have more cinematographic exposures. A complete cinematographic record of the sittings would have the advantage of determining later what had or had not taken place.[1] We should often like to know a good deal that an author does not tell us in so many words, though undoubtedly, he might have done so. For instance, A. Lehmann when referring to the experiments of Zoellner, asks with regard to the cords used during the sitting, whether Slade had had no opportunity of annexing one of them. Only an uninterrupted cinematographic film taken of the medium to include even the time outside the

[1] This could best be determined by means of stereoscopic cinematography. An apparatus of that kind should not be hard to construct.

sittings themselves could ensure a definite answer
—though this is, of course, impracticable by
reason of the expense.

Photography with ordinary light is, unfortunately, of no avail, when, as in the case of Eusapia,
the medium forbids the use of it, or when light
impedes or destroys the mediumistic phenomena.
In the future, therefore, it would be advisable to
take photographs with invisible ultra-violet
lights ; though it is, of course, impossible to
predict whether the mediumistic phenomena
will be able to stand these rays. The attempt,
however, should be made. Attempts should
also be made to keep the limbs of the medium
under constant control, by means of numerous
stereoscopic X-ray exposures. Should, however,
it be found that all these rays act destructively
on the mediumistic phenomena, it will be impossible in many cases to prove the objective
existence of the phenomena with the present
means at our disposal. We should then have
definitely to content ourselves with the reports
of witnesses based on the observations made
during the sittings by the ordinary five senses.

We cannot eliminate altogether the importance
of the question of the good faith of the person
who makes the experiments. We must accept
his word that the photographic exposures were
in order ; that the instruments were properly
installed, and that the data on the registering

apparatus tally with the curves published and have not been faked, etc.

A most interesting connexion exists between the supernormal phenomena and the spiritual worth—taken on the whole—of the life of a given individual. Many cases have been cited in both the Indian and Christian history of religion which, if true, would tend to prove that once a certain height of spiritual development is reached, parapsychic and paraphysiological phenomena are bound to follow as a matter of course. The biographies of the Indian and Christian saints are full of such happenings; they are recorded also of Christ and his Apostles in the New Testament; and the Catholic Church in consequence make canonization absolutely dependent on the testimony that such " miracles " have actually taken place. We are not yet in the position to be able to take a satisfactorily reasoned position on the question of the reality of these miracles. The whole question is closely interwoven with the other—namely to what extent parapsychic phenomena are determined by the mode of life of the individual concerned. With regard to the outstanding mediums of modern occultism, such as Helene Smith, Mrs. Piper, Eusapia Palladino, etc., it is impossible to establish any superiority of mind which would confer on them a reputation for " saintliness." But we must not lose sight of the fact that in the case of these mediums we are concerned with individuals who were born so to speak, with a parapsychically endowed personality.

The question of the connexion between parapsychic faculties and the mode of life of an individual, can only at the present time be solved in India, for modern Europe relegates ascetic " saints " to the confines of monastery or convent, where they are beyond the reach of scientific investigation.[1]

The consideration of the Indïan sphere is of further urgent necessity because of the many reports with regard to the manifestations of the mediums of that country. Unfortunately these reports so far have no claim to be considered as other than the usual travellers' tales, and, therefore, are not of sufficient value to be used as material for psychology. Further, it is not at all easy to get copies of them, and as they are mostly compiled in the English language, they mostly cannot be found at all in German libraries. Despite the urgent necessity we are still without any really scientific investigation of the Indian ascetics, fakirs, and other abnormal personalities. It is hard to understand, and regrettable in the extreme, that the Society for Psychical Research, which seems to have a special call to take the lead in this direction, has not yet

[1] Although even to-day, supernormal manifestations have been ascribed to them there.

made any effort to do so. It is obvious that the expenses entailed by a psychological expedition of such description would be considerable, but even so, much less than those of any other expedition, even on a modest scale, connected with natural science. It is no less astonishing that Indian doctors have not yet devoted themselves to the study of these problems. As universities exist in India, we might naturally have expected them to do so. Or can it be that such investigations lie hidden in Indian periodicals ? It must not be forgotten that all such investigators are, in the first instance, bound to meet with serious obstacles in getting into touch with the persons concerned. The reports of travellers often record the distrust and reserve evinced by such persons in the presence of Europeans, and their violent opposition to any prying into their secrets. Should these obstacles prove unsurmountable, it would be as well if the work were undertaken by natives who have had the benefit of a European education, and it would appear that there is no time to be lost in this respect. There is no doubt that the progress of European civilization on the one hand, and the growth of the Young India propaganda on the other, must needs diminish these manifestations to an ever greater degree, and make them of ever rarer occurrence. It gives one food for reflection when one learns that Pierre Loti, who travelled to India in order

to study the occult and theosophical under-
world and its secrets on the spot, returned to
Europe with his object unachieved, and without
having made any discovery of importance. But
despite this, it is not possible to doubt the
existence of mediums and occult circles in India,
well worthy of study by philosopher and psychologist. Certain psychic happenings, which merit
investigation, can surely be found in the circles
of Mrs. Annie Besant's " Theosophical Society."
Some isolated cases of minor significance, such as
those of the professional beggars who keep an
arm continuously outstretched before them, or
who spend their nights on a bed of thorns or
prickles, appear to be of daily and common
occurrence. And yet all closer examination of
their psychic and physical condition is lacking,
despite all the interest it would represent. We
have no opportunity in Europe for the study of
such types.

As can be seen, the field of parapsychological
problems is of the widest magnitude and importance. We are treating of discoveries which
are of equal value to the greatest discoveries of
the day in the domain of the natural science.

As a supplement to this book, I have submitted
some of the principal ideas with which we are
here concerned to closer analysis in an essay,
entitled, " Grundbegriffe der Parapsychogie "

(Fundamental conceptions of Parapsychology), Pfullingen, Baums Verlag, 1921. There is some danger of this essay being overlooked because of the rather remote place of its publication, and I think it worth while therefore to refer to it here. It is not a mere repetition of the present book, but both book and essay are complementary to each other. I have tried in the book to call attention to the reality of parapsychical phenomena and in the essay to analyze critically the resulting new conceptions.

LITERARY APPENDIX

A FEW of the more important publications on the subject are mentioned in this Appendix for the information of those interested.

The most valuable material since 1882 is contained in the " Proceedings of the Society for Psychical Research," London—a periodical which in Germany can only be obtained in the State Libraries of Berlin and Munich. This Society also publishes another journal, for circulation among its subscribers, which does not appear to be available for the general public. These publications must not be confused with the " Proceedings of the American Society for Psychical Research," which cannot be found in any German Library, and which, despite all efforts on my part, I have not been able to obtain.

F. W. Myers contributed valuable information from data which appeared in the " Proceedings for Psychical Research " up to 1905, in two volumes, published after his death, in London, 1907: " The Human Personality and its Survival after Death." The French translation, however, only comprises Myers' own text, leaving out all data from which the book emanates. So far as I know, this volume is not available at any large German Library. I have had the use of the copy belonging to the Neurobiological Institute of Berlin University. The author's theoretical point of view is that of spiritualism.

A French publication which may be compared to that of the " Proceedings " is the " Annales des Sciences Psychiques," edited by Richet. The German " Psychischen Studien " (1920, 47th year) cannot be compared to the above, as apart from some serious articles, so much that is worthless has been published in it. The occasional valuable contributions which it contains are swamped by this rubbish. This periodical must either be relieved in the future of this worthless ballast, or a purely scientific journal should be founded to take its place. This might be the " Journal für Psychologie und Neurologie " (the continuation of the Journal of Hypnotism : " Zeitschrift für Hypnotismus "), of which a considerable portion would have to be ear-marked for this purpose.

For Chapter I.—The book, " Des Indes à la Planète Mars : Etudes sur un cas de somnambulisme avec glossolalie," Geneva, 1900, 4th ed. 1909 ; first appeared after Th. Flournoy's investigations on Helene Smith. A second part followed later, entitled, " Nouvelles Observations sur un cas de Somnambulisme avec solalie," in Archives de Psychologie, Vol. 1, December, 1901—separately published at Geneva 1902. A German translation of the first book, with extracts from the second publication, appeared under the title "Die Seherin von Genf" (The Seer of Geneva), Leipzig, 1914. Second Edition, 1921, under the title, "Spiritism and Experimental Psychology."

For Chapter II.—A criticism of the investigations on Mrs. Piper is contained in a volume by Mrs. Henry Sidgwick, entitled, "A Contribution to the Study of the Psychology of Mrs. Piper's Trance Phenomena," in the Proceedings of the S.P.R., Vol. 28, 1915. The following are also of special value :—

R. Hodgson, "A Record of Observations of Certain Phenomena of Trance," Vol. 8 (1892). "A Further Record," etc., Vol. 13 (1898).

W. James, "A Record of Observations of Certain Phenomena of Trance," Part III, in Vol. 6 (1890).

W. James, "Report on Mrs. Piper's Hodgson Control," Vol. 23 (1909).

A useful insight into circumstances connected with Mrs. Piper is given in the little German book of M. Sage, "Die Mediumschaft der Frau Piper" ("The Mediumism of Mrs. Piper"), Leipzig, 1903. If a new edition of this book is brought out, it would be necessary to include the results of the medium's later development. It seems that we have now in Germany a medium like Mrs. Piper, who has been discovered and described by Dr. Joseph Böhm, of Nuremberg, in a " Collection of Essays," Pfullingen, 1921.

For Chapter III.—A whole series of articles have appeared in the S.P.R. since 1908, in connexion with the Cross-Correspondence. Alice Johnson (First Report) on "The Automatic writing of Mrs. Holland," in Vol. 21 (1909) ; Alice Johnson (Second Report), etc., in Vol. 24 (1910) ; and Third Report, etc. (Alice Johnson), in Vol. 25 (1911).

J. G. Piddington, " Three Incidents from the Sittings " (Fragment from the Report compiled by several authors : " Further Experiments with Mrs. Piper in 1918 "), in Vol. 24 (1910).

For Chapters IV and V.—W. Crookes, " Der Spiritualismus und die Wissenschaft " (Experimental Investigations on Psychic Force), German 2nd Edition, Leipzig ; also " Notes on Seances with D. D. Home," by W. Crookes. Proceedings of S.P.R., Vol. 6, 1889. Carl Friedr. Zoellner, " Wissenschaftliche Abhandlungen," Vol. 1–3, Leipzig, 1878–79 (Experiments with Slade). Fritz Grunewald, " Über eine Wiederholung des Wageversuches von Crookes," in " Psychische Studien," 1920, Books 4, 5, 8.

After the proofs had been passed I received valuable additional information which was contained in articles of Grunewald, " Physical-

Mediumistic Investigations," Pfullingen, 1920, which afford a closer insight into the modern technique for the examination of physical mediumism.

The most comprehensive volume on Eusapia Palladino is that by Enrico Morselli, entitled, " Psicologia e Spiritismo : Impressioni e note critiche sui fenomeni medianici di Eusapia Palladino," in two Vols., Turin, 1908 (with a bibliography of the entire literature written round Eusapia). The investigations of Botazzi were published in the " Annales des Sciences Psychiques," of which an extract appeared in German under the title of " The Scientific Investigations of the Phenomena of Eusapia at Naples University," by Joseph Peter, Leipzig, 1918. The report on the investigations in Paris is compiled by Jules Courtier : Rapport sur les Séances d'Eusapia Palladino a l'Institut Général Psychologique en 1915, 1916, 1917, 1918 ; Bulletin de l'Institut Général Psychologique, VIII année, Nr. 5–6, Nov., Dec., 1918. The periodical in question has long been available at the Neurobiological Institute of Berlin University, and is now obtainable in the State Library in Munich also. The official accounts of the investigations undertaken by the Committee of the Society for Psychical Research, are by : E. Feilding, W. W. Baggally, and H. Carrington, Report on a series of Sittings with Eusapia

Palladino, in the Proceedings, Vol. 23 (1909) ; E. Feilding, W. Marriott, Count Perovsky-Petrovo-Solovovo, Alice Johnson, W. W. Baggally, Report on a further series of Sittings with Eusapia at Naples, in the Proceedings, Vol. 25 (1911). Camille Flammarion, "Les Forces Naturelles Inconnues," Paris, 1907. There have also been a few observations on Slade and Eusapia Palladino by Max Dessoir, "Vom Jenseits der Seele," Stuttgart, 1917, 3rd Edition, 1920.

For Chapter V.—Works on Eva C. "Phenomena of Materialization," a Report on the investigations of mediumistic teleplasma, Munich, 1914; "The Controversy about the Phenomena of Materialization," 1914. "Sittings with Eva C.," in May and June, 1914 ; "Psychischen Studien," Vol. 41, 1914—all by A. Freiherr von Schrenck-Notzing. His book on the phenomena of materialization has now been published in English, under the title, "Phenomena of Materialization, a Contribution to the Investigation of Mediumistic Teleplasma," translated by E. E. Fournier d'Albe, London, 1920. This English edition is considerably more complete than the German original. The author enlarges in detail on the attacks made on him since his book was published, particularly with regard to the rumination hypothesis. Furthermore, the accounts of the case of Eva C. have been supplemented by reports from other investigators

down to the time of writing. The number of
illustrations, too, have been increased from 180
to 225, among which a few photographs of
materializations are included and the result of
Schrenck-Notzing's observations of an Austrian
boy on the borders of Bavaria (a case which will
surely be of great significance, as the author is
at present engaged in its close investigation).
" Les Phenomenes dits de Matérialization,"
Etude expérimentale, by Juliette Bisson, Paris,
1914; and " So-called Supernormal Physiology
and the Phenomena of Ideoplasma " (German),
Leipzig, 1920 (also " Psychischen Studien," May,
1920).

A. Freiherr von Schrenck-Notzing's, " Physical
Phenomena of Mediumism," " Studies on the
Investigation of the Processes of Telekinesis,"
Munich, refer in the first instance to the
processes of Telekinesis, as does W. J. Crawford in
his books, " The Reality of Psychic Phenomena "
(Raps, Levitation, etc.), 2nd Edition, London,
1919, and " Experiments in Psychical Science
(Levitation, Contact and the Direct Voice),
London, 1919, and " The Psychic Structures
at the Goligher Circle," London, 1921, a highly
interesting volume with numerous illustrations.
[In W. Fournier d'Albe's book, " The Goligher
Circle" (Watkins, London), the author concludes
against genuineness—Editor.] A summary, by
Schrenck-Notzing, of the twenty-six articles by

J. Ochorowicz, which were published in the "Annales des Sciences Psychiques" (vide 1st ref.), is useful also for its account of Crawford's work.

A few other recent investigations compiled in the German language with regard to other mediums are :—

"Experimental Investigations in the Domain of Clairvoyance à Distance," by A. N. Chowrin (German), Munich, 1919. "On a Case of Voluntary Clairvoyance," by W. V. Wasielewski, contained in the "Annals of Natural Philosophy," Vol. 11 (1913); also reprinted in "The Super-natural World," 24th year, 1916, No. 5. And "Telepathy and Clairvoyance: Experiments and Observations on unusual Psychic Faculties," also by W. V. Wasielewski.

"On Telepathy and Clairvoyance : Experi-mental-theoretic Investigations," by Rudolf Tischner, Munich, 1920.

"Fundamental Conceptions of Parapsycho-logy," Study on Philosophy, by T. K. Oesterreich, Pfullingen, 1920.

Alfred Lehmann's "Superstition and Witch-craft from Ancient Times Until To-day," which has been twice revised, and the enlarged edition of which was published in a German translation with references in Stuttgart, 1908, is most com-prehensive, historically speaking, and instructive in a general sense. The biased point of view of

12

the author has, however, since been superseded. His estimate of what is real must be much enlarged.

The theory of spiritism must still be looked for in the classic work on the subject, " Le Livre des Mediums," by Allan Kardec, which has appeared in very many editions. The first German translation was published under the title, " The Book of Mediums," 4th Edition, Leipzig, 1907. Carl du Prels is best informed on the subject so far as German literature is concerned, and his books, " The Riddle of Mankind," and " Spiritism," were published in the Reklams Universalbibliotek.

Amidst all that I have read in books and in innumerable communications sent to me from literature which is hard to obtain, A. N. Aksakow's chief work, " Animismus und Spiritismus," remains the most valuable. It is an attempt to test critically mediumistic phenomena with particular reference to the hypothesis of hallucination and the unconscious mind (2 Vols., 5th Edition, Leipzig, 1919). This book was written as a counter-volume to E. V. Hartmann's book, " Spiritualism," Berlin, 1885, which claimed to relegate all mediumistic phenomena—in so far as they are authentic—to the domain of telepathy and clairvoyance. In contradiction to this theory, Aksakow attempts to prove that the spiritualistic interpretation is the right one. Karl Kiesewetter's " History of the Newer

Occultism," 2nd Enlarged Edition, Leipzig, 1909, is no less valuable. The author had received many personal communications from Fechner Weber, etc.

Among the mass of spiritualistic literature, I would cite as examples the following volumes: " My Experiences in the Realm of Spiritualism," 2nd Edition, Leipzig, 1919, by M. Seiling, " The Mediumship of Mme. Elizabeth von Pribytkoff," by W. v. Pribytkoff (German), Leipzig, 1903. · " What I Have Seen," by M. T. Falkomer ; my own investigations in the bright fields of the lesser-known human faculties (German), Leipzig, 1901.

A new book, which has had a wide circulation in England, was published during the war by the physicist, Sir Oliver Lodge, whose spiritualistic tendencies are well known. The book is entitled, " Raymond, or Life and Death," London, 1916, and is mainly concerned with the automatic writings which, according to the author's conviction, emanate from his son Raymond, an English officer, who was killed in action.

" The Evidence for Communication with the Dead," by Mrs. Anna Hude, London, 1913, is of value by reason of its numerous references to the more recent English investigations. The book itself is entirely based on the results attained by the mediums, Mrs. Verrall, Mrs. Holland, and Mrs. Piper.

The Autobiography of the medium, Mrs.
d'Esperance, is entitled "In the Realm of
Shadows" (German), Berlin, 1892.

Finally, W. James' last word in regard in
Parapsychology must be included: "Final
Impressions of a Psychical Researcher," in
"Memories and Studies," London, 1911 (former-
ly published in "The American Magazine,"
October, 1909, under the title, "Confidences
of a Psychical Researcher."

Chapter VI.—The innumerable writings of
Mme. Blavatsky, Mrs. Annie Besant, as well as
Leadbeater, have all appeared in German
(Leipzig, Theosophisches Verlaghaus. The
voluminous works of Mme. Blavatsky, entitled,
"The Secret Teaching" (3 Vols.), and the
"Veiled Isis" (2 Vols.), form the basis for all
subsequent Theosophy, in so far as it is not to be
ascribed to older Indian teachings. We would
mention A. Besant's "Change in the World"
(1910), and "The Inner Life," by Leadbeater
(2 Vols., 1910), as examples. The report on
Mme. Blavatsky, by Hodgson, after his investigations in India, is published in the Proceedings of
the Society for Psychical Research, Vol. 3,
1885.

A good insight into R. Steiner's thought is
obtained by the perusal of his "Theosophie,"
"How does one Obtain Cognition of the Higher
Spheres?" "Secret Science in Outline" (all

published in Berlin, Philosophisch-Antroposo-
phischer Verlag).

Fr. Rittelmeyer's " The Life Work of Rudolf
Steiner," Munich, 1921, gives a survey of Steiner's
work.

A good description of the whole development
of modern Theosophy is to be found in L.
Johannes Frohmeyer's book on the Theosophical
movement, Stuttgart, 1920.

In conclusion, I may refer to R. Schmidt's
" Fakirs," Berlin, 1908.

" The report of the Committee of the Society
for Psychical Research will be found in its
" Proceedings," Vol. 32, January, 1922.   A very
acute criticism of this report, by Dr. G. Geley,
was published in the " Revue Metapsychique,"
1922, No. 2 (Paris).

Printed in Great Britain by Jarrold & Sons, Ltd., Norwich.

ЧР

Visit us at *www.quidprobooks.com*.

www.ingramcontent.com/pod-product-compliance
Lightning Source LLC
Chambersburg PA
CBHW070915270326
41927CB00011B/2582